Geometry

LARSON
BOSWELL
STIFF

Applying • Reasoning • Measuring

Chapter 6
Resource Book

The Resource Book contains the wide variety
of blackline masters available for Chapter 6.
The blacklines are organized by lesson. Included
are support materials for the teacher as well as
practice, activities, applications, and assessment
resources.

McDougal Littell
A HOUGHTON MIFFLIN COMPANY
Evanston, Illinois • Boston • Dallas

Contributing Authors

The authors wish to thank the following individuals for their contributions to the Chapter 6 Resource Book.

Eric J. Amendola
Patrick M. Kelly
Edward H. Kuhar
Lynn Lafferty
Dr. Frank Marzano
Wayne Nirode
Dr. Charles Redmond
Paul Ruland

ISBN: 0-618-02069-1

6789-VEI- 04 03

Contents

6 *Quadrilaterals*

Contents

Contents

Descriptions of Resources

This Chapter Resource Book is organized by lessons within the chapter in order to make your planning easier. The following materials are provided:

Tips for New Teachers These teaching notes provide both new and experienced teachers with useful teaching tips for each lesson, including tips about common errors and inclusion.

Parent Guide for Student Success This guide helps parents contribute to student success by providing an overview of the chapter along with questions and activities for parents and students to work on together.

Prerequisite Skills Review Worked-out examples are provided to review the prerequisite skills highlighted on the Study Guide page at the beginning of the chapter. Additional practice is included with each worked-out example.

Strategies for Reading Mathematics The first page teaches reading strategies to be applied to the current chapter and to later chapters. The second page is a visual glossary of key vocabulary.

Lesson Plans and Lesson Plans for Block Scheduling This planning template helps teachers select the materials they will use to teach each lesson from among the variety of materials available for the lesson. The block-scheduling version provides additional information about pacing.

Warm-Up Exercises and Daily Homework Quiz The warm-ups cover prerequisite skills that help prepare students for a given lesson. The quiz assesses students on the content of the previous lesson. (Transparencies also available)

Activity Support Masters These blackline masters make it easier for students to record their work on selected activities in the Student Edition.

Alternative Lesson Openers An engaging alternative for starting each lesson is provided from among these four types: *Application, Activity, Geometry Software,* or *Visual Approach.* (Color transparencies also available)

Technology Activities with Keystrokes Keystrokes for Geometry software and calculators are provided for each Technology Activity in the Student Edition, along with alternative Technology Activities to begin selected lessons.

Practice A, B, and C These exercises offer additional practice for the material in each lesson, including application problems. There are three levels of practice for each lesson: A (basic), B (average), and C (advanced).

Contents

Reteaching with Practice These two pages provide additional instruction, worked-out examples, and practice exercises covering the key concepts and vocabulary in each lesson.

Quick Catch-Up for Absent Students This handy form makes it easy for teachers to let students who have been absent know what to do for homework and which activities or examples were covered in class.

Cooperative Learning Activities These enrichment activities apply the math taught in the lesson in an interesting way that lends itself to group work.

Interdisciplinary Applications/Real-Life Applications Students apply the mathematics covered in each lesson to solve an interesting interdisciplinary or real-life problem.

Math and History Applications This worksheet expands upon the Math and History feature in the Student Edition.

Challenge: Skills and Applications Teachers can use these exercises to enrich or extend each lesson.

Quizzes The quizzes can be used to assess student progress on two or three lessons.

Chapter Review Games and Activities This worksheet offers fun practice at the end of the chapter and provides an alternative way to review the chapter content in preparation for the Chapter Test.

Chapter Tests A, B, and C These are tests that cover the most important skills taught in the chapter. There are three levels of test: A (basic), B (average), and C (advanced).

SAT/ACT Chapter Test This test also covers the most important skills taught in the chapter, but questions are in multiple-choice and quantitative-comparison format. (See *Alternative Assessment* for multi-step problems.)

Alternative Assessment with Rubrics and Math Journal A journal exercise has students write about the mathematics in the chapter. A multi-step problem has students apply a variety of skills from the chapter and explain their reasoning. Solutions and a 4-point rubric are included.

Project with Rubric The project allows students to delve more deeply into a problem that applies the mathematics of the chapter. Teacher's notes and a 4-point rubric are included.

Cumulative Review These practice pages help students maintain skills from the current chapter and preceding chapters.

Geometry
Chapter 6 Resource Book

LESSON 6.1

COMMON ERROR Students have a tendency to confuse the terms convex and nonconvex for polygons. Provide a few more polygonal shapes for them to copy and practice the solution method of extending the sides shown in Example 2 on page 323.

TEACHING TIP Once students know the properties of triangles they can apply them to other figures that contain triangles. This is illustrated on page 324. The quadrilateral is divided into two triangles, leading to the conclusion that the sum of the measures of the angles in a quadrilateral is 360°. Students may use this approach to divide any polygon into non-overlapping triangles and determine the sum of the measures of the interior angles.

LESSON 6.2

TEACHING TIP Demonstrate the properties illustrated in the theorems on page 330, yourself or with the aid of students, by using a large pair of identically labeled congruent parallelograms. Place them to show opposite sides congruent, opposite angles congruent, etc.

COMMON ERROR Since parallelograms have sides that slant (different from a rectangle or square), students may have trouble realizing that the diagonals are not congruent. Use the models mentioned above to show the diagonals have different lengths.

LESSON 6.3

TEACHING TIP Help students recognize that the theorems on page 338 for proving a quadrilateral is a parallelogram are the converses of those given in the previous lesson.

TEACHING TIP When using the properties to show that a quadrilateral is a parallelogram, as in Example 4 on page 341, there may be several possible approaches. Assure students that they need to show only one method of solution. However, they may use another solution method on their scrap paper as an additional verification.

COMMON ERROR When doing a problem similar to that in Example 4, students may try to show only one pair of slopes equal as in Method 1 or only one pair of sides equal as in Method 2. Caution them to make sure they are meeting the complete conditions stated in the theorems. Stress that if Method 1 is used, students must show both pairs of opposite slopes are equal. Do the same for Method 2.

LESSON 6.4

TEACHING TIP Consider having students make a flow diagram, titled parallelograms, illustrating the relationships of all the types of parallelograms to show that they understand the Venn diagram on page 347. Just as in a flow proof, a ray can connect the shapes and the definitions can be written under the name of each shape.

TEACHING TIP Discuss the coordinate diagram used in Example 5 on page 350. Students should understand why the rhombus was drawn in that location and how the coordinates were determined.

LESSON 6.5

TEACHING TIP Have the students create a flow diagram similar to the one in the previous lesson, but title this one quadrilaterals and have it include kites and trapezoids. These flow diagrams can be used as a reference when studying or reviewing the properties of the different quadrilaterals.

LESSON 6.6

INCLUSION The diagram at the top of page 364 should be similar to the flow diagram created in the previous lesson. This is a diagram that students could easily sketch for themselves on scrap paper when taking a quiz or a test. Students could also draw a similar diagram using just the general shape of the quadrilateral. It can be an excellent aid, especially for those who have limited English proficiency.

TEACHING TIP The summary charts in the exercises on page 367 can be very helpful. Consider having similar ones available for the class and let students pair up to complete them. Extra copies may be used as a summary review, a future quiz, or as an introduction to a lesson.

TEACHING TIP There are many ways to write a proof or solve problems like those in Examples 4 and 5 on page 366. Students may find that their first conjecture is not a correct one. Encourage them to draw figures if none are given, to write the given information to support the diagram, and to consider the needed properties. Then they can make and test another conjecture. This may be a time when initial errors help students to learn more and they should think of it (the process) in a positive way.

LESSON 6.7

TEACHING TIP Students should be familiar with the area formulas in the theorems on page 372 from their study of mathematics in earlier grades. However, they may not be familiar with how the formulas were developed or relate to each other. Also, knowing how all the area formulas relate to the area for a triangle can help them if they forget a formula, such as the one for area of a trapezoid.

TEACHING TIP Point out that students could find the areas for a kite or rhombus by dividing the shapes into non-overlapping triangles and using the area formula for a triangle. However, using the area formulas for a kite and a rhombus shown in the theorems on page 374 might be easier or more appropriate.

COMMON ERROR Students may forget to include the appropriate units of measure with their solutions to problems involving length and area. This is especially true when solutions involve the use of a coordinate graph. Remind students that answers are given in sentence form and must include the needed units of measure.

Outside Resources

BOOKS/PERIODICALS

Kennedy, Joe and Eric McDowell. "Geoboard Quadrilaterals." *Mathematics Teacher* (April 1998); pp. 288–290.

Housinger, Margaret M. "Trap a Surprise in an Isosceles Trapezoid." *Mathematics Teacher* (January 1996); pp. 12–14.

ACTIVITIES/MANIPULATIVES

Cuisenaire. *Geoboard Activity Mats.* White Plains, NY; Dale Seymour Publications.

SOFTWARE

Shape Makers: Developing Geometric Reasoning with the Geometer's Sketchpad. Enables students to dynamically transform a shape to explore properties of triangles and quadrilaterals. Emeryville, CA; Key Curriculum Press.

Parent Guide for Student Success

For use with Chapter 6

Chapter Overview One way that you can help your student succeed in Chapter 6 is by discussing the lesson goals in the chart below. When a lesson is completed, ask your student to interpret the lesson goals for you and to explain how the mathematics of the lesson relates to one of the key applications listed in the chart.

Lesson Title	Lesson Goals	Key Applications
6.1: Polygons	Identify, name, and describe polygons. Use the sum of the measures of the interior angles of a quadrilateral.	• Tile Patterns • Traffic Signs • Plant Shapes
6.2: Properties of Parallelograms	Use properties of parallelograms in real-life situations.	• Furniture Design • Baklava • Scissors Lift
6.3: Proving Quadrilaterals are Parallelograms	Prove that a quadrilateral is a parallelogram. Use coordinate geometry with parallelograms.	• Hinged Boxes • Bike Gears • Bird Watching
6.4: Rhombuses, Rectangles, and Squares	Use properties of sides and angles of rhombuses, rectangles, and squares. Use properties of diagonals of rhombuses, rectangles, and squares.	• Theater Set • Screen Door • Portable Table
6.5: Trapezoids and Kites	Use properties of trapezoids and kites.	• Spider Webs • Wedding Cake • Flying Kites
6.6: Special Quadrilaterals	Identify special quadrilaterals based on limited information. Prove that a quadrilateral is a special type.	• Tent Shapes • Gem Cutting
6.7: Areas of Triangles and Quadrilaterals	Find the areas of squares, rectangles, parallelograms, and triangles. Find the areas of trapezoids, kites, and rhombuses.	• Roof Area • Energy Conservation • Parade Floats

Study Strategy

Form a Study Group is the study strategy featured in Chapter 6 (see page 320). Encourage your student to form a study group with other geometry students. Having your student try to explain ideas to members of the group can provide an opportunity for him or her to pull ideas together, to identify and overcome misunderstanding, and to review and prepare for tests.

NAME _____ DATE _____

Parent Guide for Student Success

For use with Chapter 6

Key Ideas Your student can demonstrate understanding of key concepts by working through the following exercises with you.

Lesson	Exercise
6.1	If you connect the tips of the arms of a starfish, what polygon is formed? Is it regular in an "ideal" starfish?
6.2	You are making a rope ladder like the one diagrammed on the right. To make sure the rungs are horizontal, *ABCD* must be a parallelogram. Which sides are congruent? Which angles are supplementary to $\angle DAB$?
6.3	After the rope ladder in Exercise 6.2 is made, how can you make sure *ABCD* is a parallelogram? What theorem could you use?
6.4	To make the rope ladder like the one diagrammed in the Exercise for Lesson 6.2 so that the rails are vertical and the rungs are horizontal, *ABCD* would need to be a rectangle. How could you guarantee that it is a rectangle? What theorem or corollary would you use?
6.5	Quadrilateral *WXYZ* is a kite. Its diagonals intersect at point *V*. $WV = VY = 3$ m, $XV = 4$ m and $ZV = 6$ m. Find the lengths of the sides of the kite.
6.6	The diagonals of quadrilateral *JKLM* are congruent. What quadrilaterals always meet this condition? What quadrilaterals sometimes meet this condition?
6.7	Quadrilateral *ABCD* has vertices $A(-3, 2)$, $B(-1, 4)$, $C(4, 2)$, and $D(-1, 0)$. Find its area.

Home Involvement Activity

You will need: Materials to make sketches

Directions: Challenge your student to a quadrilateral scavenger hunt. Try to find and sketch a real-life example of each of the following: a square, a rectangle that is not a square, a rhombus that is not a square, a kite, a parallelogram that is not a rhombus or a rectangle, a trapezoid, and a quadrilateral that is none of these. See who can finish first.

Answers

6.1: pentagon; yes **6.2:** $\overline{AB} \cong \overline{DC}$, $\overline{AD} \cong \overline{BC}$, $\angle ADC$ and $\angle ABC$ **6.3:** *Sample answer:* make sure $AB = DC$ and $AD = BC$; Theorem 6.6 (See p. 338.) **6.4:** make sure $AC = BD$; Theorem 6.13 (See p. 349.) **6.5:** $WX = XY = 5$ m, $WZ = YZ \approx 6.71$ m **6.6:** always: rectangle, square, isosceles trapezoid; sometimes: parallelogram, rhombus **6.7:** 14 square units

Geometry
Chapter 6 Resource Book

Prerequisite Skills Review

For use before Chapter 6

EXAMPLE 1

Identifying Postulates or Theorems

Use the diagram at the right. Write the theorem or postulate that justifies the statement.

a. ∠1 and ∠2 are supplementary   **b.** ∠2 ≅ ∠3

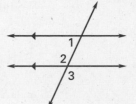

SOLUTION

a. Consecutive Interior Angles Theorem

b. Vertical Angles Theorem

Exercises for Example 1

Use the diagram at the right. Write the postulate or theorem that justifies each statement.

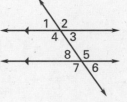

1. ∠1 ≅ ∠8   **2.** ∠2 ≅ ∠7

3. ∠3 ≅ ∠8   **4.** ∠6 supplementary to ∠7

5. ∠4 supplementary to ∠8   **6.** ∠6 ≅ ∠8

EXAMPLE 2

Writing Postulates or Theorems to Prove Triangles are Congruent

Write the postulate or theorem you could use to prove △ABC ≅ △DEF.

a. $\overline{AC} \perp \overline{BC}, \overline{DF} \perp \overline{EF}$,
∠C ≅ ∠F, $\overline{BC} \cong \overline{EF}$,
$\overline{BA} \cong \overline{ED}$

b. $\overline{AB} \cong \overline{DE}$,
∠B ≅ ∠E,
$\overline{BC} \cong \overline{EF}$

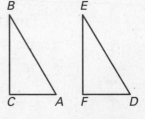

SOLUTION

a. HL Theorem

b. SAS Postulate

Exercises for Example 2

Write the postulate or theorem you could use to prove △DOG ≅ △CAT.

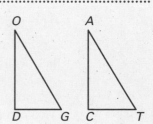

7. $\overline{DO} \cong \overline{CA}, \overline{OG} \cong \overline{AT}$, ∠O ≅ ∠A

8. $\overline{DG} \cong \overline{CT}, \overline{GO} \cong \overline{TA}, \overline{OD} \cong \overline{AC}$

9. ∠D ≅ ∠C, ∠G ≅ ∠T, $\overline{DG} \cong \overline{CT}$

10. $\overline{OD} \perp \overline{DG}, \overline{AC} \perp \overline{CT}$, ∠D ≅ ∠C, $\overline{OG} \cong \overline{AT}, \overline{DG} \cong \overline{CT}$

11. ∠G ≅ ∠T, $\overline{OG} \cong \overline{AT}, \overline{DO} \cong \overline{CA}$

12. ∠O ≅ ∠A, ∠G ≅ ∠T, $\overline{DG} \cong \overline{CT}$

NAME _____ DATE _____

Prerequisite Skills Review

EXAMPLE 3   *Finding the Length, Slope and Midpoint of a Segment*

Find the length, slope and midpoint of $\overline{AB}$.

a. $A(-2, 7)$, $B(6, 11)$ **b.** $A(4, 3)$, $B(-1, 7)$

SOLUTION

a. $AB = \sqrt{(6 - (-2))^2 + (11 - 7)^2}$ $m = \dfrac{11 - 7}{6 - (-2)}$ $M = \left(\dfrac{-2 + 6}{2}, \dfrac{7 + 11}{2}\right)$

$AB = \sqrt{8^2 + 4^2}$

$AB = \sqrt{64 + 16}$ $m = \dfrac{4}{8}$ $M = (2, 9)$

$AB = \sqrt{80}$ $m = \dfrac{1}{2}$

$AB \approx 8.94$ units

b. $AB = \sqrt{(4 - (-1))^2 + (3 - 7)^2}$ $m = \dfrac{7 - 3}{-1 - 4}$ $M = \left(\dfrac{4 + (-1)}{2}, \dfrac{3 + 7}{2}\right)$

$AB = \sqrt{5^2 + (-4)^2}$

$AB = \sqrt{25 + 16}$ $m = \dfrac{4}{-5}$ $M = \left(\dfrac{3}{2}, 5\right)$

$AB = \sqrt{41}$ $m = -\dfrac{4}{5}$

$AB \approx 6.40$ units

Exercises for Example 3

Find the length, slope and midpoint of the segment.

13. $A(9, 4)$, $B(-1, -2)$ **14.** $C(6, 5)$, $D(2, 7)$ **15.** $E(-5, 0)$, $F(0, 4)$

16. $G(8, -4)$, $H(2, -10)$ **17.** $J(0, 3)$, $K(4, 0)$ **18.** $L(-8, 4)$, $M(6, 6)$

Strategies for Reading Mathematics

For use with Chapter 6

Strategy: Reading Proofs

You will read and write many proofs in Geometry. Two types of proofs are paragraph proofs and two-column proofs. A paragraph proof is shown below.

GIVEN $\angle 1 \cong \angle 2$

PROVE $\angle ABC$ and $\angle BCD$ are supplementary.

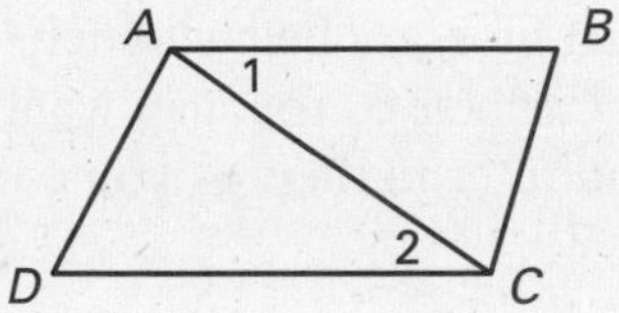

GIVEN ——

Since $\angle 1 \cong \angle 2$, $\overline{AB} \parallel \overline{DC}$ by the Alternate Interior Angles Converse Theorem. By the Consecutive Interior Angles Theorem, $\angle ABC$ and $\angle BCD$ are supplementary.

PROVE ——

STUDY TIP

Reading Paragraph and Two-Column Proofs

Always look in the first sentence or state-ment to find what is given. Always look in the last sentence or statement to find what is proved. Be sure that you understand all of the steps involved in going from what is given to what is proved.

Questions

1. What is given in the proof above? How do you know?

2. What is proved first from the given statement? How is it proved?

3. How is the final statement proved?

4. Fill in the blanks to rewrite the paragraph proof given above as a two-column proof.

Statements	Reasons
1. _____	1. _____
2. _____	2. _____
3. _____	3. _____

NAME _____ DATE _____

Strategies for Reading Mathematics

For use with Chapter 6

Visual Glossary

The Study Guide on page 320 lists the vocabulary for Chapter 6 as well as review vocabulary from previous chapters. Use the page references on page 320 or the Glossary in the textbook to review key terms from prior chapters. Use the visual glossary below to help you understand some of the key vocabulary in Chapter 6. You may want to copy these diagrams into your notebook and refer to them as you complete the chapter.

GLOSSARY

polygon (p. 322) A plane figure that meets the following two conditions. (1) It is formed by three or more segments called sides, such that no two sides with a common endpoint are collinear. (2) Each side intersects exactly two other sides, one at each endpoint.

equilateral polygon (p. 323) A polygon with all of its sides congruent.

equiangular polygon (p. 323) A polygon with all of its interior angles congruent.

regular polygon (p. 323) A polygon that is equilateral and equiangular.

parallelogram (p. 330) A quadrilateral with both pairs of opposite sides parallel.

rhombus (p. 347) A parallelogram with four congruent sides.

rectangle (p. 347) A parallelogram with four right angles.

square (p. 347) A parallelogram with four congruent sides and four right angles.

trapezoid (p. 356) A quadrilateral with exactly one pair of parallel sides.

Equilateral, Equiangular, and Regular Polygons

A polygon can be equilateral, equiangular, both, or neither. Polygons that are both equilateral and equiangular are regular.

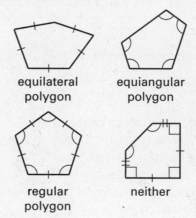

equilateral polygon   equiangular polygon

regular polygon   neither

Special Quadrilaterals

A quadrilateral is a polygon with four sides. Some special quadrilaterals are shown below.

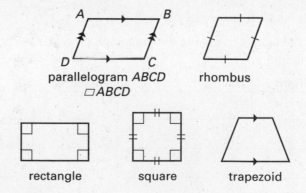

parallelogram ABCD
□ABCD

rhombus

rectangle   square   trapezoid

Geometry
Chapter 6 Resource Book

LESSON 6.1

Lesson Plan

2-day lesson (See *Pacing the Chapter*, TE pages 318C–318D)   **For use with pages 321–328**

GOALS   1. **Identify, name, and describe polygons.**
2. **Use the sum of the measures of the interior angles of a quadrilateral.**

State/Local Objectives _____

_____

✓ Check the items you wish to use for this lesson.

STARTING OPTIONS
____ Prerequisite Skills Review: CRB pages 5–6
____ Strategies for Reading Mathematics: CRB pages 7–8
____ Homework Check: TE page 305: Answer Transparencies
____ Warm-Up or Daily Homework Quiz: TE pages 322 and 308, CRB page 11, or Transparencies

TEACHING OPTIONS
____ Motivating the Lesson: TE page 323
____ Concept Activity: SE page 321
____ Lesson Opener (Application): CRB page 12 or Transparencies
____ Technology Activity with Keystrokes: CRB page 13
____ Examples: Day 1: 1–4, SE pages 322–324; Day 2: See the Extra Examples.
____ Extra Examples: Day 1 or Day 2: 1–4, TE pages 323–324 or Transp.
____ Closure Question: TE page 324
____ Guided Practice: SE page 325   Day 1: Exs. 1–11; Day 2: See Checkpoint Exs. TE pages 323–324

APPLY/HOMEWORK
Homework Assignment
____ Basic   Day 1: 12–20 even, 21–23, 24–38 even, 42–52 even; Day 2: 13–19 odd,
          25–51 odd, 54–64 even
____ Average   Day 1: 12–20 even, 21–23, 24–38 even, 42–52 even; Day 2: 13–19 odd,
          25–51 odd, 54–64 even
____ Advanced   Day 1: 12–20 even, 21–23, 24–38 even, 42–52 even; Day 2: 13–19 odd,
          25–53 odd, 54–64 even

Reteaching the Lesson
____ Practice Masters: CRB pages 14–16 (Level A, Level B, Level C)
____ Reteaching with Practice: CRB pages 17–18 or Practice Workbook with Examples
____ Personal Student Tutor

Extending the Lesson
____ Applications (Interdisciplinary): CRB page 20
____ Challenge: SE page 328; CRB page 21 or Internet

ASSESSMENT OPTIONS
____ Checkpoint Exercises: Day 1 or Day 2: TE pages 323–324 or Transp.
____ Daily Homework Quiz (6.1): TE page 328, CRB page 24, or Transparencies
____ Standardized Test Practice: SE page 328; TE page 328; STP Workbook; Transparencies

Notes _____

_____

TEACHER'S NAME _____ CLASS _____ ROOM _____ DATE _____

Lesson Plan for Block Scheduling

1-day lesson (See *Pacing the Chapter,* **TE pages 318C–318D)**   **For use with pages 321–328**

GOALS   1. **Identify, name, and describe polygons.**
2. **Use the sum of the measures of the interior angles of a quadrilateral.**

State/Local Objectives _____

_____

✓ **Check the items you wish to use for this lesson.**

STARTING OPTIONS

____ Prerequisite Skills Review: CRB pages 5–6
____ Strategies for Reading Mathematics: CRB pages 7–8
____ Homework Check: TE page 305: Answer Transparencies
____ Warm-Up or Daily Homework Quiz: TE pages 322 and
         308, CRB page 11, or Transparencies

CHAPTER PACING GUIDE	
Day	**Lesson**
1	Assess Ch. 5; **6.1 (begin)**
2	**6.1 (end)**; 6.2 (begin)
3	6.2 (end); 6.3 (begin)
4	6.3 (end); 6.4 (begin)
5	6.4 (end); 6.5 (begin)
6	6.5 (end); 6.6 (all)
7	6.7 (all)
8	Review Ch. 6; Assess Ch. 6

TEACHING OPTIONS

____ Motivating the Lesson: TE page 323
____ Concept Activity: SE page 321
____ Lesson Opener (Application): CRB page 12 or Transparencies
____ Technology Activity with Keystrokes: CRB page 13
____ Examples: Day 1: 1–4, SE pages 322–324; Day 2: See the Extra Examples.
____ Extra Examples: Day 1 or Day 2: 1–4, TE pages 323–324 or Transp.
____ Closure Question: TE page 324
____ Guided Practice: SE page 325   Day 1: Exs. 1–11; Day 2: See Checkpoint Exs. TE pages 323–324

APPLY/HOMEWORK

Homework Assignment (See also the assignment for Lesson 6.2.)
____ Block Schedule: Day 1: 12–20 even, 21–23, 24–38 even, 42–52 even; Day 2: 13–19 odd,
         25–51 odd, 54–64 even

Reteaching the Lesson
____ Practice Masters: CRB pages 14–16 (Level A, Level B, Level C)
____ Reteaching with Practice: CRB pages 17–18 or Practice Workbook with Examples
____ Personal Student Tutor

Extending the Lesson
____ Applications (Interdisciplinary): CRB page 20
____ Challenge: SE page 328; CRB page 21 or Internet

ASSESSMENT OPTIONS

____ Checkpoint Exercises: Day 1 or Day 2: TE pages 323–324 or Transp.
____ Daily Homework Quiz (6.1): TE page 328, CRB page 24, or Transparencies
____ Standardized Test Practice: SE page 328; TE page 328; STP Workbook; Transparencies

Notes _____

_____

_____

NAME _____ DATE _____

WARM-UP EXERCISES

For use before Lesson 6.1, pages 321–328

1. What is the sum of the measures of the interior angles of a triangle?

2. Two angles in a triangle measure 34° and 53°. The third angle measures $3x°$. What is x?

3. In which kind of triangle are all three sides congruent?

4. In which kind of triangle are all three angles congruent?

....

DAILY HOMEWORK QUIZ

For use after Lesson 5.6, pages 302–308

In Exercises 1–4, complete with <, >, or =.

1. AB _____ DE

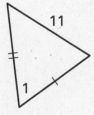 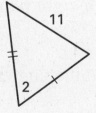

2. $m\angle 1$ _____ $m\angle 2$

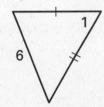 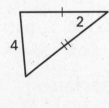

3. $m\angle 1$ _____ $m\angle 2$

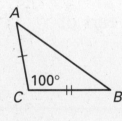 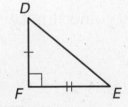

4. AB _____ CD

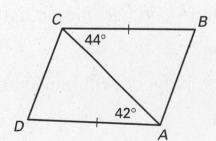

NAME _____ DATE _____

Application Lesson Opener

For use with pages 322–328

Use the prefix chart below and the given definition to write a definition of the boldfaced word.

tri-	3	quadr-	4	penta-	5	hexa-	6
hepta-	7	oct-	8	nona-	9	dec-	10

1. triplet: One of three children born at one birth.

    quadruplet

2. quadrennial: Happening once every four years.

    octennial

3. decathalon: An athletic contest that consists of ten events for each participant.

    pentathalon

4. hexapod: Having six legs or feet.

    tripod

5. octogenarian: A person between eighty and ninety years of age.

    nonagenarian

6. pentagram: A five-pointed star.

    hexagram

7. octahedron: A solid geometric figure with eight plane faces.

    decahedron

8. nonagon: A polygon with nine sides and nine angles.

    heptagon

9. heptameter: A line of verse consisting of seven metrical feet.

    pentameter

Geometry
Chapter 6 Resource Book

NAME _____ DATE _____

Technology Activity Keystrokes

For use with page 327

Keystrokes for Exercise 40

TI-92

1. Draw quadrilateral *ABCD*.

   F3 4 (Place cursor at desired location for point *A*.) **ENTER** *A* (Place cursor at desired location for point *B*.) **ENTER** *B* (Place cursor at desired location for point *C*.) **ENTER** *C* (Place cursor at desired location for point *D*.) **ENTER** *D* (Move cursor to point *A*.) **ENTER**

2. Measure ∠*DAB*, ∠*ABC*, ∠*BCD*, and ∠*CDA*.

   F6 3 (Place cursor on point *D*.) **ENTER** (Place cursor on point *A*.) **ENTER** (Move cursor to point *B*.) **ENTER**

   Repeat for the other angles.

3. Calculate the sum of the four angles.

   F6 6 (Move cursor and highlight angle *DAB*.) **ENTER** **+** (Move cursor to highlight angle *ABC*.) **ENTER** **+** (Move cursor to highlight angle *BCD*.) **ENTER** **+** (Move cursor to highlight angle *CDA*.) **ENTER** **ENTER**

   (The result will appear on the screen.)

4. Drag a vertex of the quadrilateral.

   F1 1 (Place cursor on point *A*.) **ENTER**

   (Use the drag key [☞] and the cursor pad to drag the point.)

SKETCHPAD

1. Draw quadrilateral *ABCD*. Select segment from the straightedge tools and make four segments for quadrilateral *ABCD*.

2. Measure ∠*DAB*, ∠*ABC*, ∠*BCD*, and ∠*CDA*. To measure angle *DAB*, choose the selection arrow tool, select point *D*, hold the shift key down, and select points *A* and *B*. Then choose **Angle** from the **Measure** menu. Repeat for the remaining angles. Before selecting the next angle, be sure to click anywhere in the work area to deselect the previous points.

3. Calculate the sum of the four angles. Select **Calculate** from the **Measure** menu. Click each angle measure, insert **+** sign between each measure, and click OK.

4. Use the translate selection arrow tool to drag a vertex of the quadrilateral.

Geometry
Chapter 6 Resource Book

Decide whether the figure is a polygon. If not, explain why.

1.
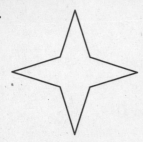

2.
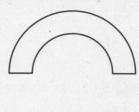

3.
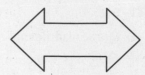

Use the number of sides to tell what kind of polygon the shape is. Then state whether the polygon is *convex* or *concave*.

4.
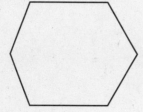

5.

6.
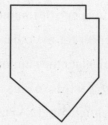

Use the diagram at the right to answer the following.

7. Name the polygon by the number of sides it has.

8. Polygon *MNOPQR* is one name for the polygon. State two other names.

9. Name all of the diagonals that have vertex *M* as an endpoint.

10. Name the consecutive angles to ∠*N*.

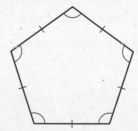

State whether the polygon is best described as *equilateral*, *equiangular*, *regular*, or *none of these*.

11.
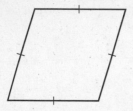

12.
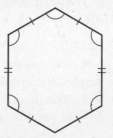

13.
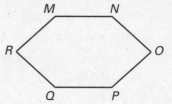

Use the information in the diagram to solve for *x*.

14.
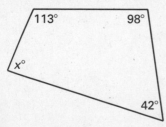

15.
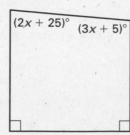

16.
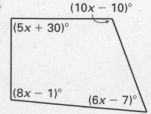

NAME _____   DATE _____

Practice B

For use with pages 322–328

Decide whether the figure is a polygon. If not, explain why.

1.
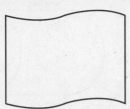

2.
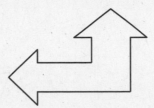

3.

Use the number of sides to tell what kind of polygon the shape is. Then state whether the polygon is *convex* or *concave*.

4.
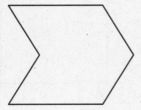

5.
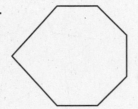

6.
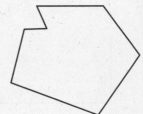

Use the diagram at the right to answer the following.

7. Name the polygon by the number of sides it has.

8. Polygon *ABCDEFG* is one name for the polygon.
   State two other names.

9. Name all of the diagonals that have vertex *E* as an endpoint.

10. Name the nonconsecutive angles to ∠*A*.

Draw a figure that fits the description.

11. A convex octagon

12. A concave decagon

13. An equilateral quadrilateral that is not equiangular

14. An equiangular quadrilateral that is not equilateral

15. An equiangular hexagon that is not regular

Use the information in the diagram to solve for *x*.

16.
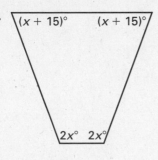

$(x + 15)°$ $(x + 15)°$

$2x°$ $2x°$

17.

$(3x - 1)°$
$(5x - 15)°$
$(4x + 34)°$
$(5x - 15)°$

18.
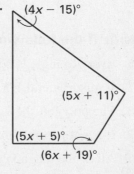

$(4x - 15)°$
$(5x + 11)°$
$(5x + 5)°$
$(6x + 19)°$

Practice C

For use with pages 322–328

Decide whether the figure is a polygon. If not, explain why.

1.

2.

3.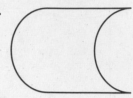

Use the number of sides to tell what kind of polygon the shape is. Then state whether the polygon is *convex* or *concave*.

4.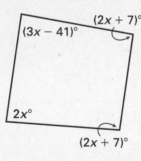

5.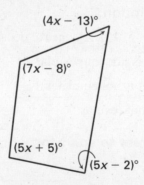

6.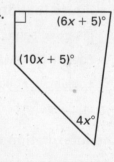

Draw a figure that fits the description.

7. A convex hexagon

8. A concave nonagon

9. An equilateral hexagon that is not equiangular

10. An equiangular hexagon that is not equilateral

11. An equiangular quadrilateral that is not regular

Use the information in the diagram to solve for *x*.

12.
$(3x - 41)°$ $(2x + 7)°$
$2x°$
$(2x + 7)°$

13.
$(4x - 13)°$
$(7x - 8)°$
$(5x + 5)°$
$(5x - 2)°$

14.
$(6x + 5)°$
$(10x + 5)°$
$4x°$

Decide if the following statements are *true* or *false*.

15. Every triangle is convex.

16. $\overline{BE}$ is a diagonal of polygon *BACDE*.

17. If quadrilateral *WXYZ* is regular, then it has four right angles.

18. The polygon shown in Exercise 5 is a regular polygon.

19. It is not possible to draw a concave quadrilateral.

20. All of the diagonals of a regular polygon are congruent.

NAME _____ DATE _____

Reteaching with Practice

For use with pages 322–328

GOAL Identify, name, and describe polygons and use the sum of the measures of the interior angles of a quadrilateral

---

VOCABULARY

A **polygon** is a plane figure that is formed by three or more segments called sides, such that no two sides with a common endpoint are collinear, and each side intersects exactly two other sides, one at each endpoint. Each endpoint of a side is a **vertex** of the polygon.

A polygon is **convex** if no line that contains a side of the polygon contains a point in the interior of the polygon.

A polygon that is not convex is called **nonconvex** or **concave.**

A **diagonal** of a polygon is a segment that joins two *nonconsecutive* vertices.

Theorem 6.1   Interior Angles of a Quadrilateral
The sum of the measures of the interior angles of a quadrilateral is 360°.

EXAMPLE 1 *Identifying Polygons*

State whether the figure is a polygon. If it is not, explain why.

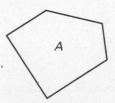

SOLUTION

Figures *A* and *C* are polygons.

- Figure *B is not* a polygon because it only has two sides, and one of its sides is not a segment.

- Figure *D is not* a polygon because two of the sides intersect only one other side.

Exercises for Example 1

State whether each figure is a polygon. If it is not, explain why.

1.   2.   3.   4.

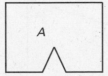   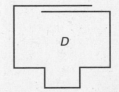

NAME _____ DATE _____

Reteaching with Practice

For use with pages 322–328

EXAMPLE 2 ## Identifying Convex and Concave Polygons

State whether each polygon is convex or concave.

a.

b.

SOLUTION

a. The polygon has 5 sides. When extended, none of the sides intersect the interior, so the polygon is convex.

b. The polygon has 10 sides. When extended, some of the sides intersect the interior, so the polygon is concave.

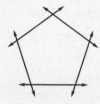

Exercises for Example 2

State whether the polygon is *convex* or *concave*.

5.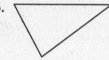

6.

7.

EXAMPLE 3 ## Interior Angles of a Quadrilateral

Find $m\angle A$ and $m\angle B$.

SOLUTION

Find the value of x. Use the Interior Angles of a Quadrilateral Theorem to write an equation involving x. Then solve the equation.

$5x° + 7x° + 50° + 70° = 360°$ Theorem 6.1

$x = 20$ Solve for x.

So, $m\angle A = 5x° = 5(20)° = 100°$ and $m\angle B = 7x° = 7(20)° = 140°$.

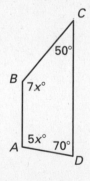

Exercises for Example 3

Use the information in the diagram to solve for x.

8.

9.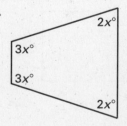

NAME ———————————————————————————————— DATE —————————

Quick Catch-Up for Absent Students

For use with pages 321–328

The items checked below were covered in class on (date missed) ———————

Activity 6.1: Classifying Shapes (p. 321)

———— **Goal:** Determine which shapes are polygons.

Lesson 6.1: Polygons

———— **Goal 1:** Identify, name, and describe polygons. (pp. 322–323)

Material Covered:

———— Student Help: Study Tip

———— Example 1: Identifying Polygons

———— Example 2: Identifying Convex and Concave Polygons

———— Example 3: Identifying Regular Polygons

Vocabulary:

polygon, p. 322   sides, p. 322

vertex, p. 322   convex, p. 323

nonconvex, p. 323   concave, p. 323

equilateral, p. 323   equiangular, p. 323

regular, p. 323

———— **Goal 2:** Use the sum of the measures of the interior angles of a quadrilateral. (p. 324)

Material Covered:

———— Student Help: Study Tip

———— Example 4: Interior Angles of a Quadrilateral

Vocabulary:

diagonal, p. 324

———— Other (specify) ————————————————————————————————————

——

Homework and Additional Learning Support

———— Textbook (specify) _pp. 325–328_————————————————————————

——

———— *Reteaching with Practice* worksheet (specify exercises)————————————

———— *Personal Student Tutor* for Lesson 6.1

Interdisciplinary Application

For use with pages 322–328

Algebraic Formula for the Angles of a Polygon

ALGEBRA   From your work with triangles, you already know that the sum of the measures of the interior angles of any triangle equals 180°. Theorem 6.1 states that the sum of the interior angles of a quadrilateral is 360°. What if you wanted to know the sum of the interior angle measures for a 100-gon? The process of writing a formula that will allow you to answer this question often begins with finding a pattern. Patterns are an invaluable tool in mathematics, but you would not want to write out a pattern 100 times! Algebra enables you to use the pattern to write a formula to find the sum of the interior angles of any polygon if you know the number of sides.

1. Copy and complete the table using a pattern or measurement.

Polygon	*Number of sides*	*Sum of the interior angles*
Triangle	3	180°
Quadrilateral	4	360°
Pentagon	?	?
Hexagon	?	?
Heptagon	?	?
Octagon	?	?

2. As the number of sides of a polygon increases by one, what happens to the total measure of the interior angles?

3. Write an algebraic formula to find the sum of the measures of the interior angles of any polygon given the number of sides.

4. Using your formula from Exercise 3, verify your answers in Exercise 1.

5. Find the sum of the measures of the interior angles of a 100-gon using the algebraic formula from Exercise 3.

Lesson 6.1

NAME _____ DATE _____

Challenge: Skills and Applications

For use with pages 322–328

Use the information in the diagram to solve for *x*. Diagrams may not be drawn to scale.

1.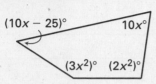
$(10x - 25)°$ $10x°$
$(3x^2)°$ $(2x^2)°$

2.
$(x^2 - 5x)°$
$(x^2)°$
$(x^2 + 2x)°$ $30°$

3.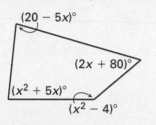
$(20 - 5x)°$
$(2x + 80)°$
$(x^2 + 5x)°$
$(x^2 - 4)°$

4.
$(100 - 3x)°$
$(80 - 4x)°$
$(120 + 2x)°$
$(60 + 5x)°$

Use the information in the diagram to solve for *x* and *y*.

5.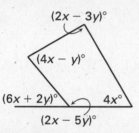
$(2x - 3y)°$
$(4x - y)°$
$(6x + 2y)°$ $4x°$
$(2x - 5y)°$

6.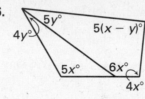
$5y°$
$4y°$ $5(x - y)°$
$5x°$ $6x°$
$4x°$

Use the diagram to find the sum of the interior angles in the polygon.

7. Pentagon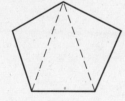

8. Hexagon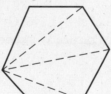

9. Octagon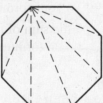

Tell whether the statement is *true* or *false*. If it is false, sketch a counterexample.

10. Every equiangular polygon is convex.

11. Every equilateral polygon is convex.

TEACHER'S NAME _____ CLASS _____ ROOM _____ DATE _____

Lesson Plan

2-day lesson (See *Pacing the Chapter,* TE pages 318C–318D)   **For use with pages 329–337**

GOALS 1. **Use some properties of parallelograms.**
          2. **Use properties of parallelograms in real-life situations.**

State/Local Objectives _____

_____

✓ **Check the items you wish to use for this lesson.**

STARTING OPTIONS
____ Homework Check: TE page 325: Answer Transparencies
____ Warm-Up or Daily Homework Quiz: TE pages 330 and 328, CRB page 24, or Transparencies

TEACHING OPTIONS
____ Motivating the Lesson: TE page 331
____ Lesson Opener (Activity): CRB page 25 or Transparencies
____ Technology Activity with Keystrokes: CRB pages 26–27
____ Examples: Day 1: 1–6, SE pages 331–333; Day 2: See the Extra Examples.
____ Extra Examples: Day 1 or Day 2: 1–6, TE pages 331–333 or Transp.
____ Technology Activity: SE page 329
____ Closure Question: TE page 333
____ Guided Practice: SE page 333   Day 1: Exs. 1–19; Day 2: See Checkpoint Exs. TE pages 331–333

APPLY/HOMEWORK
Homework Assignment
____ Basic   Day 1: 20–38 even, 40–44, 56, 58; Day 2: 21–39 odd, 45–50, 55, 57, 60, 61, 66–74 even
____ Average   Day 1: 20–38 even, 40–44, 56, 58; Day 2: 21–39 odd, 45–55, 57, 60, 61, 66–74 even
____ Advanced   Day 1: 20–38 even, 40–44, 56, 58; Day 2: 21–39 odd, 45–55, 57, 60–64, 66–74 even

Reteaching the Lesson
____ Practice Masters: CRB pages 28–30 (Level A, Level B, Level C)
____ Reteaching with Practice: CRB pages 31–32 or Practice Workbook with Examples
____ Personal Student Tutor

Extending the Lesson
____ Applications (Real-Life): CRB page 34
____ Challenge: SE page 337; CRB page 35 or Internet

ASSESSMENT OPTIONS
____ Checkpoint Exercises: Day 1 or Day 2: TE pages 331–333 or Transp.
____ Daily Homework Quiz (6.2): TE page 337, CRB page 38, or Transparencies
____ Standardized Test Practice: SE page 337; TE page 337; STP Workbook; Transparencies

Notes _____

_____

_____

TEACHER'S NAME _____ CLASS _____ ROOM _____ DATE _____

Lesson Plan for Block Scheduling

1-day lesson (See *Pacing the Chapter*, TE pages 318C–318D)   For use with pages 329–337

GOALS   1. **Use some properties of parallelograms.**
            2. **Use properties of parallelograms in real-life situations.**

State/Local Objectives _____

_____

_____

✓ **Check the items you wish to use for this lesson.**

STARTING OPTIONS

____ Homework Check: TE page 325: Answer Transparencies
____ Warm-Up or Daily Homework Quiz: TE pages 330 and
       328, CRB page 24, or Transparencies

TEACHING OPTIONS

____ Motivating the Lesson: TE page 331
____ Lesson Opener (Activity): CRB page 25 or Transparencies
____ Technology Activity with Keystrokes: CRB pages 26–27
____ Examples: Day 2: 1–6, SE pages 331–333; Day 3: See the Extra Examples.
____ Extra Examples: Day 2 or Day 3: 1–6, TE pages 331–333 or Transp.
____ Technology Activity: SE page 329
____ Closure Question: TE page 333
____ Guided Practice: SE page 333   Day 2: Exs. 1–19; Day 3: See Checkpoint Exs. TE pages 331–333

APPLY/HOMEWORK

Homework Assignment (See also the assignments for Lessons 6.1 and 6.3.)
____ Block Schedule: Day 2: 20–40 even, 41–44, 56, 58; Day 3: 21–39 odd, 45–55, 57, 60, 66–74 even

Reteaching the Lesson
____ Practice Masters: CRB pages 28–30 (Level A, Level B, Level C)
____ Reteaching with Practice: CRB pages 31–32 or Practice Workbook with Examples
____ Personal Student Tutor

Extending the Lesson
____ Applications (Real-Life): CRB page 34
____ Challenge: SE page 337; CRB page 35 or Internet

ASSESSMENT OPTIONS

____ Checkpoint Exercises: Day 2 or Day 3: TE pages 331–333 or Transp.
____ Daily Homework Quiz (6.2): TE page 337, CRB page 38, or Transparencies
____ Standardized Test Practice: SE page 337; TE page 337; STP Workbook; Transparencies

CHAPTER PACING GUIDE	
Day	**Lesson**
1	Assess Ch. 5; 6.1 (begin)
2	6.1 (end); **6.2 (begin)**
3	**6.2 (end)**; 6.3 (begin)
4	6.3 (end); 6.4 (begin)
5	6.4 (end); 6.5 (begin)
6	6.5 (end); 6.6 (all)
7	6.7 (all)
8	Review Ch. 6; Assess Ch. 6

Lesson 6.2

Notes _____

_____

_____

NAME _____ DATE _____

WARM-UP EXERCISES

For use before Lesson 6.2, pages 329–337

Which property justifies the statement?

1. If $\angle A \cong \angle B$ and $\angle B \cong \angle C$, then $\angle A \cong \angle C$.

2. If $\angle M \cong \angle N$, then $\angle N \cong \angle M$.

3. If $m\angle 1 = 80°$, and $m\angle 1 = m\angle 2$, then $m\angle 2 = 80°$.

DAILY HOMEWORK QUIZ

For use after Lesson 6.1, pages 321–328

1. Use the number of sides to identify
   the polygon. Is it *convex* or *concave*?

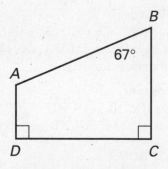

2. Draw an equilateral quadrilateral that
   is not regular.

3. Use the information in the diagram
   to find $m\angle A$.

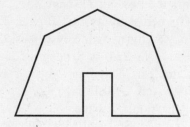

4. Use the information in the diagram
   to solve for x.

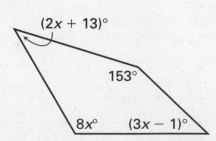

NAME _____ DATE _____

Activity Lesson Opener

For use with pages 330–337

SET UP: Work in a group.
YOU WILL NEED: • **flat toothpicks** • **ruler**

1. Break some toothpicks so that your group has about 8 different lengths to use. Keep some of the toothpicks full length. Mark the midpoint of each toothpick. Place a toothpick flat on a piece of paper and draw a point on the paper at each endpoint. Label the points *A* and *B*. Then place another toothpick over the first one, at any angle, with the midpoints aligned. Draw a point on the paper at each endpoint of the second toothpick, and label the points *Y* and *Z*. Remove the toothpicks. Draw figure *AYBZ*.

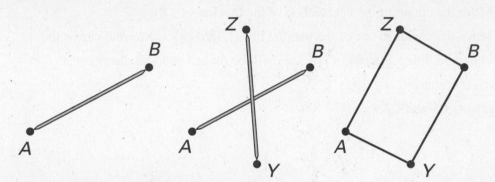

2. Repeat Exercise 1 until you have various sizes and shapes for figure *AYBZ*. How can you change the size of *AYBZ*? How can you change the shape of *AYBZ* without changing toothpicks? Describe the general shape of *AYBZ*.

3. On one of your figures *AYBZ*, draw $\overline{AB}$ and $\overline{YZ}$. How were these lengths determined? What are these segments called? What is true about $\overline{AB}$ and $\overline{YZ}$ for every figure *AYBZ*?

4. Make some conjectures about the properties of figure *AYBZ*. Consider opposite sides, opposite angles, consecutive angles, and diagonals. Write the name of figure *AYBZ* if you know it.

Technology Activity Keystrokes

For use with page 329

TI-92

Construct

1. Draw $\overline{AB}$.

   [F2] 5 (Move cursor to desired location for point *A*.) [ENTER] *A* (Move cursor to desired location for point *B*.) [ENTER] *B*

   Use the same feature to draw $\overline{AC}$.

2. Construct a line through point *B* parallel to $\overline{AC}$.

   [F4] 2 (Place the cursor on point *B*.) [ENTER] (Move the cursor to $\overline{AC}$.) [ENTER]

3. Construct a line through point *C* parallel to $\overline{AB}$.

   [F4] 2 (Place the cursor on point *C*.) [ENTER] (Move the cursor to $\overline{AB}$.) [ENTER]

4. Label the intersection of the two lines as point *F* and then hide the lines.

   [F2] 3 (Move the cursor to the intersection of the two lines.) [ENTER] *F*

   [F7] 1 (Move the cursor to one of the parallel lines.) [ENTER] (Move the cursor to the second parallel line.) [ENTER] (The lines will be dashed and will disappear when the next command is given.)

5. Draw segments *BF* and *CF*.

Investigate

1. Drag point *A*.

   [F1] 1 (Place cursor on point *A*.) [ENTER] (Use the drag key 🖑 and the cursor pad to drag the point.) Repeat for points *B* and *C*.

2. Measure $\overline{AB}$, $\overline{BF}$, $\overline{CF}$, and $\overline{AC}$.

   [F6] 1 (Move cursor to $\overline{AB}$.) [ENTER] (Move cursor to $\overline{BF}$.) [ENTER] (Move cursor to $\overline{CF}$.) [ENTER] (Move cursor to $\overline{AC}$.) [ENTER]

3. See Investigate Step 1.

5. Measure $\angle A$, $\angle B$, $\angle C$, and $\angle F$.

   [F6] 3 (Place cursor on point *B*.) [ENTER] (Place cursor on point *A*.) [ENTER] (Place cursor on point *C*.) [ENTER] Repeat for the other angles.

   Drag point *A*.

   [F1] 1 (Place cursor on point *A*.) (Use the drag key 🖑 and the cursor pad to drag the point.)

Technology Activity Keystrokes

For use with page 329

Extension

Draw diagonals $\overline{AF}$ and $\overline{BC}$ using the segment command. (**F2** 5)

Label the intersection point of the two diagonals as point G using the intersection point command. (**F2** 3)

Measure $\overline{AB}$, $\overline{BG}$, $\overline{FG}$, and $\overline{CG}$.

F6 1 (Place cursor on point A.) **ENTER** (Move cursor to point B.) **ENTER**

Repeat for the other segments.

SKETCHPAD

Construct

1. Draw $\overline{AB}$ and $\overline{AC}$. Choose segment from the straightedge tools.

2. Construct a line through point B parallel to $\overline{AC}$. Use the selection arrow tool to select $\overline{AC}$ and point B. Choose **Parallel Line** from the **Construct** menu.

3. Construct a line through point C parallel to $\overline{AB}$. See Step 2.

4. Label the intersection of the two lines as point F and then hide the lines. Select point from the toolbox and click on the intersection point. Relabel the point. Use the selection arrow tool to select the two lines. Select **Hide Lines** from the **Display** menu.

5. Draw $\overline{BF}$ and $\overline{CF}$. Select segment from the straightedge tools.

Investigate

1. Use the translate selection arrow tool to drag points A, B, and C.

2. Measure $\overline{AB}$, $\overline{BF}$, $\overline{CF}$, and $\overline{AC}$. Use the selection arrow tool to select the four segments. Choose **Length** from the **Measure** menu.

3. Use the translate selection arrow tool to drag points A, B, or C.

5. Measure $\angle A$, $\angle B$, $\angle C$, and $\angle F$. To measure $\angle A$, use the selection arrow tool to select points B, A, and C. Then select **Angle** from the **Measure** menu. Repeat for the remaining angles. Before selecting the next angle, be sure to click anywhere in the work area to deselect the previous points. Use the translate selection arrow tool to drag points A, B, and C.

Extension

Draw diagonals $\overline{AF}$ and $\overline{BC}$. Choose segment from the straightedge tools. Label the intersection point of the two diagonals as point G. Select point from the toolbox and click on the intersection point.

Measure $\overline{AB}$, $\overline{BG}$, $\overline{FG}$, and $\overline{CG}$. Use the selection arrow tool to select the four segments. Choose **Length** from the **Measure** menu.

Practice A

For use with pages 330–337

Decide whether the figure is a parallelogram. If it is not, explain why not.

1.
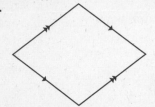

2.
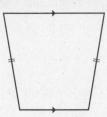

3.
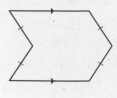

Use the diagram of parallelogram *MNOP* at the right. Complete the statement, and give a reason for your answer.

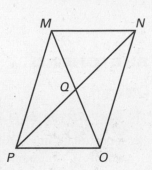

4. $\overline{MN} \cong$ ___?___

5. $\overline{MN} \parallel$ ___?___

6. $\overline{ON} \cong$ ___?___

7. $\angle MPO \cong$ ___?___

8. $\overline{PQ} \cong$ ___?___

8. $\overline{QM} \cong$ ___?___

10. $\angle MQN \cong$ ___?___

11. $\angle NPO \cong$ ___?___

Find the measure in the parallelogram *HIJK*. Explain your reasoning.

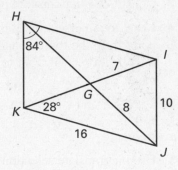

12. HI

13. KH

14. GH

15. HJ

16. $m\angle KIH$

17. $m\angle JIH$

18. $m\angle KJI$

19. $m\angle HKI$

Find the value of each variable in the parallelogram.

20.
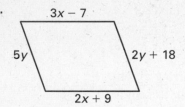

21.
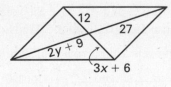

Complete the flow-proof at the right.

22. **Given:** $\square ABCD$

   Prove: $\triangle ABD \cong \triangle CDB$

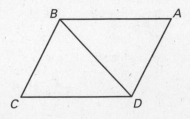

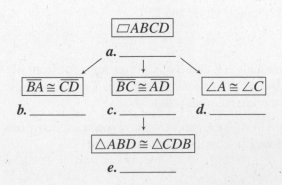

$\square ABCD$

a. _____

$\overline{BA} \cong \overline{CD}$ $\overline{BC} \cong \overline{AD}$ $\angle A \cong \angle C$

b. _____   c. _____   d. _____

$\triangle ABD \cong \triangle CDB$

e. _____

Lesson 6.2

Practice B

For use with pages 330–337

Decide whether the figure is a parallelogram. If it is not, explain why not.

1.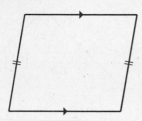

2.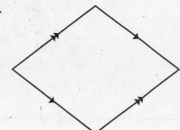

3.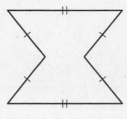

Use the diagram of parallelogram *KLMN* at the right. Points *O, P, Q, R* are midpoints of $\overline{XN}$, $\overline{XK}$, $\overline{XL}$, and $\overline{XM}$. Find the indicated measures.

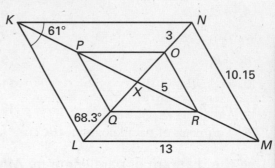

4. *KN*   5. *KL*

6. *XN*   7. *LN*

8. *KP*   9. *KR*

10. $m\angle MNL$ 11. $m\angle NLM$

12. $m\angle NML$ 13. $m\angle XQP$

14. Perimeter of parallelogram *KLMN*

Find the value of each variable in the parallelogram.

15.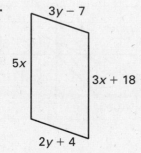

16.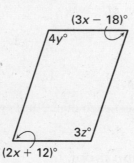

17.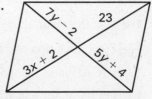

Write a two-column or a paragraph proof.

18. **Given:** ▱*ABCD*

    Prove: $\triangle AED \cong \triangle CEB$

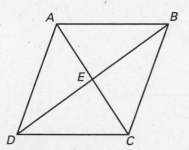

19. **Given:** ▱*WXYZ*

    $\overline{ZM} \perp \overline{WX}$, $\overline{XN} \perp \overline{ZY}$

    Prove: $\triangle ZMW \cong \triangle XNY$

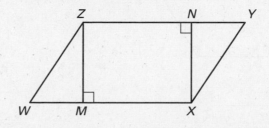

Practice C

For use with pages 330–337

Decide whether the figure is a parallelogram. If it is not, explain why not.

1.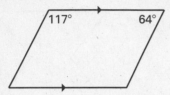

2.

3.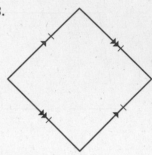

Use the diagram of parallelogram *ABCD* at the right. $\overline{AC} \perp \overline{BD}$.
Find the indicated measures.

4. *AE*

5. *AD*

6. *EB*

7. *DB*

8. *AB*

9. Perimeter △*AEB*

10. *m* ∠*DBA*

11. *m* ∠*DEC*

12. *m* ∠*ACD*

13. *m* ∠*CAB*

14. Perimeter of parallelogram *ABCD*

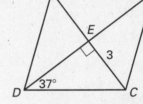

Use the diagram of parallelogram *MNOP* at the right.

15. Use the Distance Formula to show $\overline{MP} \cong \overline{NO}$.

16. Use the Distance Formula to show $\overline{MN} \cong \overline{PO}$.

17. Find the slope of $\overline{MP}$ and $\overline{NO}$.

18. Are $\overline{MP}$ and $\overline{NO}$ parallel? Explain.

19. Do the diagonals $\overline{MO}$ and $\overline{NP}$ bisect each other? Justify your answer.

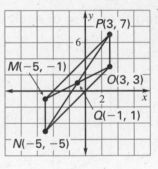

Write a two-column or a paragraph proof.

20. **Given:** □*MATH*
    $\overline{MN} \cong \overline{AT}$
    Prove: ∠1 ≅ ∠2

21 **Given:** □*ATRO*
    $\overline{PT} \cong \overline{IP}$
    Prove: ∠*I* ≅ ∠*AOR*

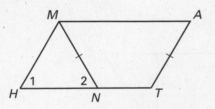

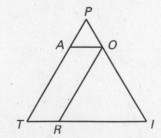

Reteaching with Practice

For use with pages 330–337

GOAL **Use some properties of parallelograms**

> **VOCABULARY**
>
> A **parallelogram** is a quadrilateral with both pairs of opposite sides parallel.
>
> **Theorem 6.2**
> If a quadrilateral is a parallelogram, then its opposite sides are congruent.
>
> **Theorem 6.3**
> If a quadrilateral is a parallelogram, then its opposite angles are congruent.
>
> **Theorem 6.4**
> If a quadrilateral is a parallelogram, then its consecutive angles are supplementary.
>
> **Theorem 6.5**
> If a quadrilateral is a parallelogram, then its diagonals bisect each other.

EXAMPLE 1 *Using Properties of Parallelograms*

ABCD is a parallelogram. Find the lengths and angle measures.

   a. *AD*   **b.** *EC*

   c. $m\angle ADC$ **d.** $m\angle BCD$

SOLUTION

   a. $AD = BC$ from Theorem 6.2. So, $AD = 8$.

   b. From Theorem 6.5, the two diagonals of *ABCD* bisect each other. Therefore, $AE = EC$. So, $EC = 5$.

   c. $m\angle ABC = m\angle ADC$ from Theorem 6.3.
       $m\angle ABC = m\angle ABE + m\angle CBE$ by the Angle Addition Postulate.
       Substituting, $m\angle ADC = 65° + 45° = 110°$.

   d. $m\angle BCD + m\angle ADC = 180°$ by Theorem 6.4. So
       $m\angle BCD = 180° - m\angle ADC$ by the Subtraction Property of
       Equality. By substituting and simplifying, $m\angle BCD = 70°$.

Exercises for Example 1

Find the value of each variable in the parallelogram.

1.

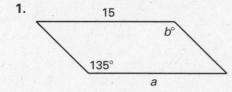

2.

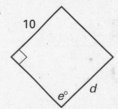

Lesson 6.2

NAME _____ DATE _____

Reteaching with Practice

For use with pages 330–337

3.

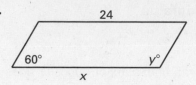

24

60°

$y°$

x

4.

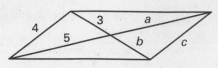

4   3 a

5   b c

5.

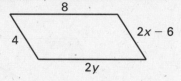

$e°$ $f°$

82° $d°$

6.

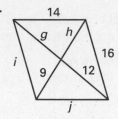

14

g h

i 16

9   12

j

EXAMPLE 2 **Using Algebra with Parallelograms**

Use algebra to find the value of each variable in the parallelogram.

a.

8

4

$2x - 6$

$2y$

b.

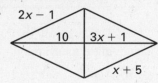

$2x - 1$

10    $3x + 1$

$x + 5$

SOLUTION

Set up equations based upon the properties of parallelograms provided in Theorems 6.2 through 6.5.

a. Because opposite sides of a parallelogram are congruent (Theorem 6.2), $2x - 6 = 4$. Solving for x yields $2x = 10$ which means $x = 5$. Also by Theorem 6.2, $2y = 8$, so $y = 4$.

b. From Theorem 6.2, $2x - 1 = x + 5$. Thus, $x = 6$.
From Theorem 6.5, $3x + 1 = 10$. Thus, $3x = 9$ which means $x = 3$.

Exercises for Example 2

Find the value of each variable in the parallelogram.

7.

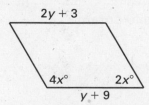

$2y + 3$

$4x°$ $2x°$

$y + 9$

8.

$(3x - 9)°$

$2y + 21$

$(2x + 31)°$

$4y + 5$

9.

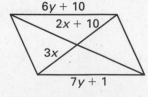

$6y + 10$

$2x + 10$

$3x$

$7y + 1$

Lesson 6.2

NAME _____ DATE _____

Quick Catch-Up for Absent Students

For use with pages 329–337

The items checked below were covered in class on (date missed) _____

Activity 6.2: Investigating Parallelograms (p. 329)

____ **Goal:** Use geometry software to explore the properties of parallelograms.

____ Student Help: Software Help

Lesson 6.2: Properties of Parallelograms

____ **Goal 1:** Use some properties of parallelograms. (pp. 330–331)

Material Covered:

____ Example 1: Using Properties of Parallelograms

____ Example 2: Using Properties of Parallelograms

____ Example 3: Using Algebra with Parallelograms

Vocabulary:

parallelogram, p. 330

____ **Goal 2:** Use properties of parallelograms in real-life situations. (pp. 332–333)

Material Covered:

____ Example 4: Proving Facts about Parallelograms

____ Example 5: Proving Theorem 6.2

____ Example 6: Using Parallelograms in Real Life

____ Other (specify) _____

_____

Homework and Additional Learning Support

____ Textbook (specify) _pp. 333–337_____

_____

____ *Reteaching with Practice* worksheet (specify exercises)_____

____ *Personal Student Tutor* for Lesson 6.2

NAME _____ DATE _____

Real-Life Application: When Will I Ever Use This?

For use with pages 330–337

Washington, D.C.

Washington, D.C. became the capital of the United States in 1800 and has since grown into a major metropolis visited by millions of people each year. It is home to major federal government buildings, such as the Capitol, and the White House, not to mention historical sites like the Washington Monument and the Lincoln Memorial.

In 1791, President George Washington chose the site for the new capital on the banks of the Potomac River and appointed a French engineer named Pierre Charles L'Enfant to design the city. With the help of American surveyors Andrew Ellicot and Benjamin Banneker, Washington, D.C. became one of the few cities in the world to actually be planned before being built.

Today, with millions of people living in the metropolitan area, efficient transportation is a major challenge. The street map below of a portion of Washington, D.C. shows how many geometrical concepts go into planning streets and highways. Although most of the streets form a rectangle, notice the parallelogram that is not a rectangle.

In Exercises 1–4, refer to the map of Washington, D.C. below. Assume that *ABCD* is a parallelogram. Justify your answer to the question using the theorems about parallelograms.

1. The length of sides $\overline{AB}$ and $\overline{BC}$ is 5 centimeters. Find *CD* and *DA*.

2. The measure of $\angle A$ is 110°. What is the measure of $\angle C$?

3. Find the measure of $\angle D$.

4. Is *ABCD* a square? Why or why not?

5. What is an advantage of having streets that form nonrectangular parallelograms?

Challenge: Skills and Applications

For use with pages 330–337

In Exercises 1–5, assume *PQRS* is a parallelogram.

1. If $PQ = x^2 - 10$ and $SR = 3x$, find all possible values of x.

2. If $PQ = 9 - x^2$ and $QR = x + 2$, find all possible values of x.

3. If $m\angle P = (x^2)°$ and $m\angle Q = 11x°$, find all possible values of x.

4. If $m\angle Q = 5x°$, $m\angle R = (3x - 2y)°$, and $m\angle S = (3x + 5y)°$, find all possible values of x and y.

5. If $m\angle P = (8y + 2)°$, $m\angle R = (y^2 - 18)°$, and $m\angle S = (2x^2)°$, find all possible values of x and y.

6. Refer to the diagram. Write a two-column proof.

   Given: *IJKL* is a parallelogram whose
   diagonals intersect at *O*.

   Prove: *O* is the midpoint of $\overline{MN}$.

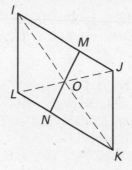

7. Refer to the diagram. Write a two-column proof.

   Given: *ABCD* and *EFGH* are parallelograms.

   Prove: $\triangle FAE \cong \triangle HCG$

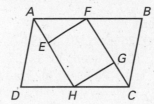

8. Two sides of a parallelogram are shown.
   Use a compass and straightedge to construct
   the remaining two sides.

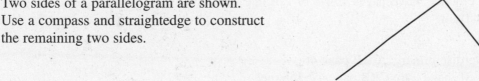

TEACHER'S NAME _____ CLASS _____ ROOM _____ DATE _____

Lesson Plan

2-day lesson (See *Pacing the Chapter,* TE pages 318C–318D)   For use with pages 338–346

 GOALS   1. **Prove that a quadrilateral is a parallelogram.**
2. **Use coordinate geometry with parallelograms.**

State/Local Objectives _____

_____

✓ **Check the items you wish to use for this lesson.**

STARTING OPTIONS
____ Homework Check: TE page 334: Answer Transparencies
____ Warm-Up or Daily Homework Quiz: TE pages 338 and 337, CRB page 38, or Transparencies

TEACHING OPTIONS
____ Lesson Opener (Activity): CRB page 39 or Transparencies
____ Examples: Day 1: 1–4, SE pages 339–341; Day 2: See the Extra Examples.
____ Extra Examples: Day 1 or Day 2: 1–4, TE pages 339–341 or Transp.; Internet
____ Closure Question: TE page 341
____ Guided Practice: SE page 342   Day 1: Exs. 1–8; Day 2: See Checkpoint Exs. TE pages 339–341

APPLY/HOMEWORK
Homework Assignment
____ Basic   Day 1: 10–26 even, 30, 32, 34–37; Day 2: 9–25 odd, 29, 40, 42–47; Quiz 1: 1–4
____ Average   Day 1: 10–26 even, 30, 32, 34–37; Day 2: 9–31 odd, 40, 42–47; Quiz 1: 1–4
____ Advanced   Day 1: 10–26 even, 30, 32, 34–37; Day 2: 9–31 odd, 38, 40, 42–47; Quiz 1: 1–4

Reteaching the Lesson
____ Practice Masters: CRB pages 40–42 (Level A, Level B, Level C)
____ Reteaching with Practice: CRB pages 43–44 or Practice Workbook with Examples
____ Personal Student Tutor

Extending the Lesson
____ Applications (Interdisciplinary): CRB page 46
____ Math & History: SE page 346; CRB page 47; Internet
____ Challenge: SE page 345; CRB page 48 or Internet

ASSESSMENT OPTIONS
____ Checkpoint Exercises: Day 1 or Day 2: TE pages 339–341 or Transp.
____ Daily Homework Quiz (6.3): TE page 345, CRB page 52, or Transparencies
____ Standardized Test Practice: SE page 345; TE page 345; STP Workbook; Transparencies
____ Quiz (6.1–6.3); SE page 346; CRB page 49

Notes _____

_____

_____

LESSON 6.3

Lesson Plan for Block Scheduling

1-day lesson (See *Pacing the Chapter,* TE pages 318C–318D)   **For use with pages 338–346**

GOALS   1. **Prove that a quadrilateral is a parallelogram.**
2. **Use coordinate geometry with parallelograms.**

State/Local Objectives _____

_____

_____

✓ Check the items you wish to use for this lesson.

STARTING OPTIONS

____ Homework Check: TE page 334: Answer Transparencies
____ Warm-Up or Daily Homework Quiz: TE pages 338 and
     337, CRB page 38, or Transparencies

TEACHING OPTIONS

____ Lesson Opener (Activity): CRB page 39 or Transparencies
____ Examples: Day 3: 1–4, SE pages 339–341; Day 4: See the Extra Examples.
____ Extra Examples: Day 3 or Day 4: 1–4, TE pages 339–341 or Transp.; Internet
____ Closure Question: TE page 341
____ Guided Practice: SE page 342   Day 3: Exs. 1–8; Day 4: See Checkpoint Exs. TE pages 339–341

APPLY/HOMEWORK
Homework Assignment (See also the assignments for Lessons 6.2 and 6.4.)
____ Block Schedule: Day 3: 10–26 even, 30, 32, 34–37; Day 4: 9–31 odd, 40, 42–47; Quiz 1: 1–4

Reteaching the Lesson
____ Practice Masters: CRB pages 40–42 (Level A, Level B, Level C)
____ Reteaching with Practice: CRB pages 43–44 or Practice Workbook with Examples
____ Personal Student Tutor

Extending the Lesson
____ Applications (Interdisciplinary): CRB page 46
____ Math & History: SE page 346; CRB page 47; Internet
____ Challenge: SE page 345; CRB page 48 or Internet

ASSESSMENT OPTIONS
____ Checkpoint Exercises: Day 3 or Day 4: TE pages 339–341 or Transp.
____ Daily Homework Quiz (6.3): TE page 345, CRB page 52, or Transparencies
____ Standardized Test Practice: SE page 345; TE page 345; STP Workbook; Transparencies
____ Quiz (6.1–6.3); SE page 346; CRB page 49

CHAPTER PACING GUIDE	
Day	**Lesson**
1	Assess Ch. 5; 6.1 (begin)
2	6.1 (end); 6.2 (begin)
3	6.2 (end); **6.3 (begin)**
4	**6.3 (end)**; 6.4 (begin)
5	6.4 (end); 6.5 (begin)
6	6.5 (end); 6.6 (all)
7	6.7 (all)
8	Review Ch. 6; Assess Ch. 6

Notes _____

_____

_____

Lesson 6.3

WARM-UP EXERCISES

For use before Lesson 6.3, pages 338–346

Give the definition, theorem, or postulate that justifies the statement.

1. If $\overline{AC} \cong \overline{A'C'}$, $\overline{AB} \cong \overline{A'B'}$, and $\overline{BC} \cong \overline{B'C'}$, then $\triangle ABC \cong \triangle A'B'C'$.

2. If $ABCD$ is a parallelogram, then $\overline{AB} \cong \overline{DC}$ and $\overline{AD} \cong \overline{BC}$.

3. If $MNPQ$ is a parallelogram, then $\overline{MP}$ bisects $\overline{NQ}$.

DAILY HOMEWORK QUIZ

For use after Lesson 6.2, pages 329–337

1. Find the sum of the lengths of the diagonals in parallelogram *FGHJ*. Explain your reasoning.

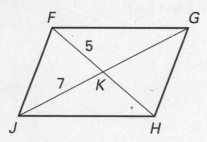

2. Each quadrilateral in the figure is a parallelogram. What is the relationship of $\overline{AB}$ and $\overline{EF}$? Explain.

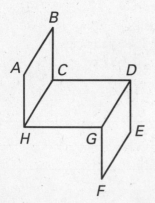

Geometry
Chapter 6 Resource Book

NAME _____ DATE _____

Activity Lesson Opener

For use with pages 338–346

SET UP: Work with a partner.

Is *ABCD* a parallelogram? To decide, one partner draws and labels figure *ABCD* with the given properties while the other partner tries to draw a figure *ABCD* that has the given properties but is *not* a parallelogram. When *ABCD* is not a parallelogram, include a sketch that supports your answer. Switch roles for each exercise.

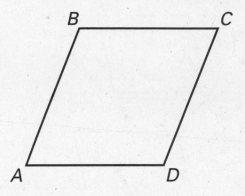

1. $\overline{AB} \cong \overline{CD}$ and $\overline{BC} \cong \overline{AD}$

2. $\overline{BD}$ and $\overline{AC}$ bisect each other.

3. $\overline{BD}$ and $\overline{AC}$ are angle bisectors.

4. $\overline{AB} \parallel \overline{CD}$ and $\overline{BC} \cong \overline{AD}$

5. $\overline{BC} \cong \overline{AD}$ and $\overline{BC} \parallel \overline{AD}$

6. $\angle A \cong \angle C$ and $\angle B \cong \angle D$

7. $\angle A$ and $\angle B$ are supplementary; $\angle C$ and $\angle D$ are supplementary.

8. $\angle A$ is supplementary to $\angle B$ and $\angle D$.

9. $\overline{BD} \cong \overline{AC}$

10. $\overline{AB} \parallel \overline{CD}$ and $\overline{BC} \parallel \overline{AD}$

NAME _____ DATE _____

Practice A

For use with pages 338–346

Are you given enough information to determine whether the quadrilateral is a parallelogram? Explain.

1.
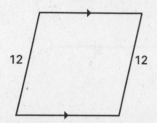

2.
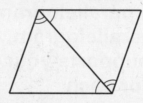

3.
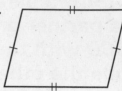

4.
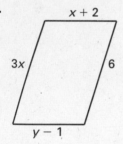

5.
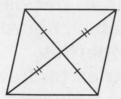

6.
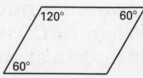

What additional information is needed in order to prove that quadrilateral *ABCD* is a parallelogram?

7. $\overline{AB} \parallel \overline{DC}$

8. $\overline{AB} \cong \overline{DC}$

9. $\angle DCA \cong \angle BAC$

10. $\overline{DE} \cong \overline{EB}$

11. $m\angle CDA + m\angle DAB = 180°$

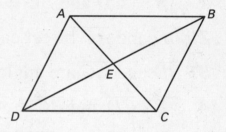

What value of *x* and *y* will make the polygon a parallelogram?

12.
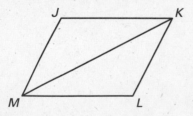

13.
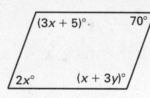

14.
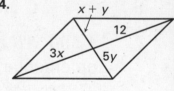

Write a two-column or a paragraph proof using each method.

15. **Given:** $\triangle MJK \cong \triangle KLM$

    Prove: *MJKL* is a parallelogram.

a. By Theorem 6.6: If both pairs of opposite sides of a quadrilateral are congruent, then the quadrilateral is a parallelogram.

b. By Theorem 6.10: If one pair of opposite sides of a quadrilateral are congruent and parallel, then the quadrilateral is a parallelogram.

Practice B

For use with pages 338–346

Decide whether each piece of given information alone is sufficient to prove that quadrilateral *ABCD* is a parallelogram.

1. E is the midpoint of $\overline{AC}$ and $\overline{BD}$.

2. $m\angle ABC + m\angle BCD = 180°$

3. $\overline{AB} \parallel \overline{DC}$ and $\overline{BC} \cong \overline{DA}$

4. $\angle ABC \cong \angle ADC$, and $\angle BAD \cong \angle BCD$

5. $\triangle ABE \cong \triangle DCE$

6. $\triangle ABE \cong \triangle CDE$

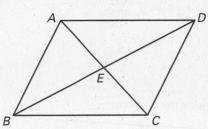

What value of *x* and *y* will make the polygon a parallelogram?

7.

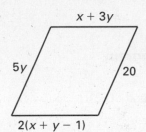

x + 3y
5y
20
2(x + y − 1)

8.

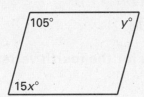

105°   y°
15x°

9.

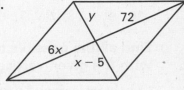

y   72
6x
x − 5

Prove that the points represent the vertices of a parallelogram. Use a different method for each exercise.

10. $A(2, -1)$, $B(1, 3)$, $C(6, 5)$, and $D(7, 1)$

11. $A(-2, -4)$, $B(1, 2)$, $C(2, 10)$, and $D(-1, 4)$

Use the diagram of the adjustable hat rack at the right to answer the following.

12. Draw the quadrilateral *ABCD*.

13. If the hat rack were expanded outward, would *ABCD* still be a parallelogram? Explain.

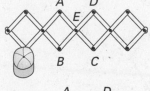

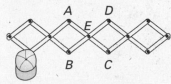

Write a two-column or a paragraph proof.

14. **Given:** $\overline{AB} \cong \overline{CD}$, $\overline{BC} \cong \overline{AF}$

$\angle AFD \cong \angle ADF$

Prove: *ABCD* is a parallelogram.

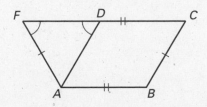

15 **Given:** $\triangle RQP \cong \triangle ONP$

R is the midpoint of $\overline{MQ}$.

Prove: *MRON* is a parallelogram.

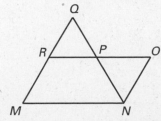

NAME _____ DATE _____

Practice C
For use with pages 338–346

Decide whether you are given enough information to determine that the quadrilateral is a parallelogram.

1. Opposite sides are parallel.

2. Opposite sides are congruent.

3. Two pairs of consecutive sides are congruent.

4. Two pairs of consecutive angles are congruent.

5. Diagonals are congruent.

6. Diagonals bisect each other.

7. All four sides are congruent.

8. Consecutive angles are supplementary.

Prove that the points represent the vertices of a parallelogram. Use a different method for each exercise.

9. $A(-4, 7)$, $B(3, 0)$, $C(2, -5)$, and $D(-5, 2)$

10. $A(-2, 8)$, $B(2, 7)$, $C(5, 1)$, and $D(1, 2)$

Find all the possible coordinates for the fourth vertex of a parallelogram with the given vertices.

11. $(4, -1)$, $(-4, 1)$, and $(0, 8)$

12. $(3, -4)$, $(-2, -1)$, and $(1, 2)$

Write a two-column or a paragraph proof.

13. **Given:** Regular hexagon $JKLMNO$

　　Prove: $OKLN$ is a parallelogram.

14. **Given:** $VWKJ$ and $SJRU$ are parallelograms.

　　Prove: $\angle W \cong \angle U$

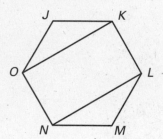

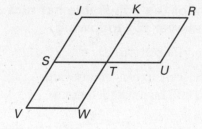

15. **Given:** $\square ABCD$

　　　　E is the midpoint of $\overline{AD}$.

　　　　F is the midpoint of $\overline{BC}$.

　　Prove: Quadrilateral $ABFE$ is a parallelogram.

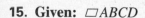

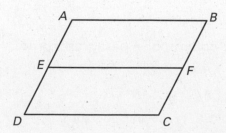

NAME _____ DATE _____

Reteaching with Practice

For use with pages 338–346

GOAL **Prove that a quadrilateral is a parallelogram and use coordinate geometry with parallelograms**

Theorem 6.6
If both pairs of opposite sides of a quadrilateral are congruent, then the quadrilateral is a parallelogram.

Theorem 6.7
If both pairs of opposite angles of a quadrilateral are congruent, then the quadrilateral is a parallelogram.

Theorem 6.8
If an angle of a quadrilateral is supplementary to both of its consecutive angles, then the quadrilateral is a parallelogram.

Theorem 6.9
If the diagonals of a quadrilateral bisect each other, then the quadrilateral is a parallelogram.

Theorem 6.10
If one pair of opposite sides of a quadrilateral are congruent and parallel, then the quadrilateral is a parallelogram.

Ways to Prove a Shape is a Parallelogram

• Show that both pairs of opposite sides are parallel.

• Show that both pairs of opposite sides are congruent.

• Show that both pairs of opposite angles are congruent.

• Show that one angle is supplementary to both consecutive angles.

• Show that the diagonals bisect each other.

• Show that one pair of opposite sides are congruent and parallel.

EXAMPLE 1 *Using Properties of Parallelograms*

Show that $A(2, 0)$, $B(3, 4)$, $C(-2, 6)$, and $D(-3, 2)$ are the vertices of a parallelogram.

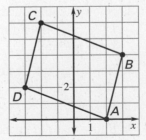

SOLUTION

There are many ways to solve this problem.

Method 1 Show that opposite sides have the same slope, so they are parallel.

$$\text{Slope of } \overline{AB} = \frac{4 - 0}{3 - 2} = 4$$

$$\text{Slope of } \overline{CD} = \frac{2 - 6}{-3 - (-2)} = \frac{-4}{-1} = 4$$

Lesson 6.3

NAME _____ DATE _____

Reteaching with Practice

For use with pages 338–346

$$\text{Slope of } \overline{BC} = \frac{6 - 4}{-2 - 3} = \frac{2}{-5} = -\frac{2}{5}$$

$$\text{Slope of } \overline{DA} = \frac{0 - 2}{2 - (-3)} = \frac{-2}{5} = -\frac{2}{5}$$

$\overline{AB}$ and $\overline{CD}$ have the same slope, so they are parallel. Similarly, $\overline{BC} \parallel \overline{DA}$. Because opposite sides are parallel, $ABCD$ is a parallelogram.

Method 2   Show that the opposite sides have the same length.

$$AB = \sqrt{(3 - 2)^2 + (4 - 0)^2} = \sqrt{17}$$

$$CD = \sqrt{(-3 - (-2))^2 + (2 - 6)^2} = \sqrt{17}$$

$$BC = \sqrt{(-2 - 3)^2 + (6 - 4)^2} = \sqrt{29}$$

$$DA = \sqrt{(2 - (-3))^2 + (0 - 2)^2} = \sqrt{29}$$

$\overline{AB} \cong \overline{CD}$ and $\overline{BC} \cong \overline{DA}$. Because both pairs of opposite sides are congruent, $ABCD$ is a parallelogram.

Method 3   Show that one pair of opposite sides is congruent and parallel. Find the slopes and lengths of $\overline{AB}$ and $\overline{CD}$ as shown in Methods 1 and 2.

$$\text{Slope of } \overline{AB} = \text{Slope of } \overline{CD} = 4$$

$$AB = CD = \sqrt{17}$$

$\overline{AB}$ and $\overline{CD}$ are congruent and parallel, so $ABCD$ is a parallelogram.

Exercises for Example 1

Refer to the methods demonstrated in Example 1 to show that the quadrilateral with the given vertices is a parallelogram.

1. Show that the quadrilateral with vertices $A(-3, 0)$, $B(-2, -4)$, $C(-7, -6)$, and $D(-8, -2)$ is a parallelogram using Method 1 from Example 1.

2. Show that the quadrilateral with vertices $A(-4, 1)$, $B(1, 2)$, $C(4, -4)$, and $D(-1, -5)$ is a parallelogram using Method 2 from Example 1.

3. Show that the quadrilateral with vertices $A(0, -6)$, $B(4, -5)$, $C(6, 3)$, and $D(2, 2)$ is a parallelogram using Method 3 from Example 1.

4. Show that the quadrilateral with vertices $A(-1, -2)$, $B(5, -3)$, $C(6, 6)$, and $D(0, 7)$ is a parallelogram using any of the three methods demonstrated in Example 1.

NAME _____ DATE _____

Quick Catch-Up for Absent Students

For use with pages 338–346

The items checked below were covered in class on (date missed) _____

Lesson 6.3: Proving Quadrilaterals are Parallelograms

____ **Goal 1:** Prove that a quadrilateral is a parallelogram. (pp. 338–340)

Material Covered:

____ Activity: Investigating Properties of Parallelograms

____ Example 1: Proof of Theorem 6.6

____ Example 2: Proving Quadrilaterals are Parallelograms

____ Example 3: Proof of Theorem 6.10

____ **Goal 2:** Use coordinate geometry with parallelograms. (p. 341)

Material Covered:

____ Student Help: Study Tip

____ Example 4: Using Properties of Parallelograms

____ Other (specify) _____

_____

Homework and Additional Learning Support

____ Textbook (specify) pp. 342–346 _____

_____

____ Internet: Extra Examples at www.mcdougallittell.com

____ *Reteaching with Practice* worksheet (specify exercises)_____

____ *Personal Student Tutor* for Lesson 6.3

NAME _____ DATE _____

Interdisciplinary Application

For use with pages 338-346

Louvre

HISTORY   Located on the Seine River in Paris and originally built as residence for the royalty of France, the Louvre now is home to one of the most renowned art collections in the world. Among its 8 miles of displays and more than a million pieces of art, such famous works as the *Venus de Milo* and the *Mona Lisa* can be found.

The Louvre has gone through many phases of construction throughout the centuries, beginning in about 1200 and taking its present appearance in the 1850s. A major project to expand and modernize the Louvre was started in 1984 under the direction of an American architect I.M. Pei. Visitors to the museum now enter through a contemporary glass pyramid, which is an engineering feat in itself.

The Pyramide Inversee, or Inverted Pyramid, has four surfaces made entirely of glass. Each surface is a combination of 21 glass quadrilaterals and 7 glass triangles.

In Exercises 1 and 2, use the figure below that models one of the glass quadrilaterals from a lateral surface of the pyramid.

1. Copy and complete the proof.

   Given: $\overline{AE} \cong \overline{EC}, \overline{BE} \cong \overline{ED}$

   Prove: *ABCD* is a parallelogram.

   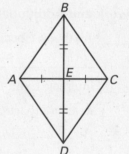

Statements	Reasons
1. $\overline{AE} \cong \overline{EC}, \overline{BE} \cong \overline{ED}$	1. _____
2. *ABCD* is a parallelogram.	2. _____

2. Using the same given information from Exercise 1, prove that *ABCD* is a parallelogram using a different theorem.

NAME _____ DATE _____

Math and History Application

For use with page 346

HISTORY   Some of the earliest recorded mathematics that exists today dates back to ancient Egypt (3000 B.C. to A.D. 260). Egyptian mathematics was written on *papyrus*, a paper-like substance made from the papyrus plant, which grows along the Nile River. Because of Egypt's dry climate, the writings are well preserved.

Most of what we know about Egyptian mathematics comes from Napoleon's unsuccessful 1798 invasion of Egypt. In addition to 38,000 soldiers, Napoleon also brought several scholars with him to research various aspects of Egyptian society. The result was the discovery of an advanced civilization that existed long before the ancient Greek and Roman empires.

Ancient Egyptian mathematics was motivated by necessity. Each spring, the Nile River would overflow, forcing farmers to reset the boundaries between their properties. In an area where usable soil was scarce, this was no small matter. It was for this reason that Egyptian scholars developed ways for calculating the area of two-dimensional regions.

MATH   One Egyptian papyrus gives instructions for finding the area of a circle. Consider a circle inscribed in a 3-by-3 grid (see figure at the left below). In the figure at the right below, an octagon is drawn.

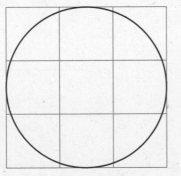   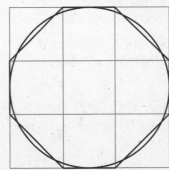

        Figure 1   **Figure 2**

1. Explain how you can use the second figure to approximate the area of the circle. Use your method to approximate the circle's area.

2. Using the modern formula for the area of a circle $A = \pi r^2$, show that the Egyptian method gives the approximation $\pi \approx 3.111\ldots\ldots$

Challenge: Skills and Applications

For use with pages 338–346

1. Refer to the diagram. Write a two-column proof.

   Given: *UWXZ* is a parallelogram, $\angle 1 \cong \angle 8$.

   Prove: *UVXY* is a parallelogram.

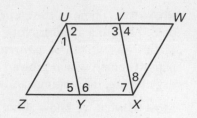

2. Refer to the diagram. Write a two-column proof.

   Given: *GIJL* is a parallelogram.

   Prove: *HIKL* is a parallelogram.

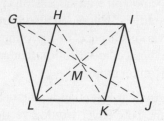

3. Try to prove the following conjecture. If you cannot prove it, explain what goes wrong when you attempt to prove it.

   If a quadrilateral has a pair of congruent opposite sides and a pair of congruent opposite angles, then it is a parallelogram.

4. In the diagram, *ABCD* is a parallelogram. Use a compass and straightedge to complete the following construction.

   a. Sketch an arc of radius *AC*, centered at *A*. Let *E* be the point (other than *C*) where this arc intersects $\overrightarrow{BC}$.

   b. Sketch an arc of radius *AD*, centered at *A*. Let *F* be the point (near *D*) on this arc so that $m\angle EAF = m\angle CAD$.

   c. Draw $\triangle AEF$.

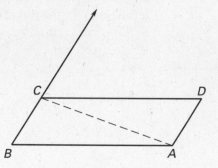

5. Refer to the figure you constructed in Exercise 4.

   a. Prove that $\overline{AB} \cong \overline{FE}$ and $\angle ABE \cong \angle AFE$.

   b. Is the conjecture in Exercise 3 *true* or *false*? Explain why.

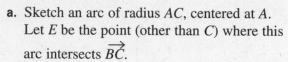
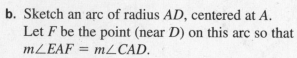
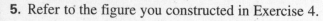

NAME _____ DATE _____

Quiz 1

For use after Lessons 6.1–6.3

1. Choose the words that describe the quadrilateral at the right: *concave*, *convex*, *equilateral*, *equiangular*, and *regular*. (*Lesson 6.1*)

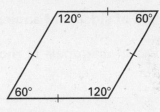

Answers

1. _____

   _____

   _____

2. _____

2. Find the value of y. Explain your reasoning. (*Lesson 6.1*)

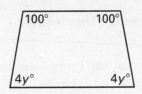

   _____

   _____

   _____

   _____

   _____

   _____

   _____

3. Write a Proof. (*Lesson 6.2*)

   Given: $m\angle 1 = 90°$
   $\overline{AB} \parallel \overline{CE} \parallel \overline{GF}$; $\overline{AC} \parallel \overline{BG} \parallel \overline{EF}$

   Prove: $\angle A$ and $\angle F$ are right angles.

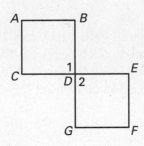

3. _____

   _____

   _____

   _____

   _____

   _____

4. Describe two ways to show that $X(0, 0)$, $Y(5, 0)$, $Z(7, 5)$, and $W(2, 5)$ are the vertices of a parallelogram. (*Lesson 6.3*)

   _____

   _____

4. _____

   _____

   _____

   _____

   _____

TEACHER'S NAME _____ CLASS _____ ROOM _____ DATE _____

Lesson Plan

2-day lesson (See *Pacing the Chapter*, TE pages 318C–318D)   For use with pages 347–355

GOALS   1. **Use properties of sides and angles of rhombuses, rectangles, and squares.**
2. **Use properties of diagonals of rhombuses, rectangles, and squares.**

State/Local Objectives _____

_____

✓ Check the items you wish to use for this lesson.

STARTING OPTIONS
_____ Homework Check: TE page 342: Answer Transparencies
_____ Warm-Up or Daily Homework Quiz: TE pages 347 and 345, CRB page 52, or Transparencies

TEACHING OPTIONS
_____ Motivating the Lesson: TE page 348
_____ Lesson Opener (Geometry Software): CRB page 53 or Transparencies
_____ Technology Activity with Keystrokes: CRB pages 54–57
_____ Examples: Day 1: 1–3, SE pages 347–348; Day 2: 4–6, SE pages 349–350
_____ Extra Examples: Day 1: TE page 348 or Transp.; Day 2: TE pages 349–350 or Transp.; Internet
_____ Closure Question: TE page 350
_____ Guided Practice: SE page 351   Day 1: Exs. 1, 3–8, 10–11; Day 2: Exs. 2, 9

APPLY/HOMEWORK
Homework Assignment
_____ Basic   Day 1: 12–43; Day 2: 44–60, 66–68, 74–82 even, 83
_____ Average   Day 1: 12–43; Day 2: 44–62, 66–68, 74–82 even, 83
_____ Advanced   Day 1: 12–43; Day 2: 44–62, 66–72, 74–82 even, 83

Reteaching the Lesson
_____ Practice Masters: CRB pages 58–60 (Level A, Level B, Level C)
_____ Reteaching with Practice: CRB pages 61–62 or Practice Workbook with Examples
_____ Personal Student Tutor

Extending the Lesson
_____ Applications (Real-Life): CRB page 64
_____ Challenge: SE page 355; CRB page 65 or Internet

ASSESSMENT OPTIONS
_____ Checkpoint Exercises: Day 1: TE page 348 or Transp.; Day 2: TE pages 349–350 or Transp.
_____ Daily Homework Quiz (6.4): TE page 355, CRB page 68, or Transparencies
_____ Standardized Test Practice: SE page 355; TE page 355; STP Workbook; Transparencies

Notes _____

_____

_____

TEACHER'S NAME _____ CLASS _____ ROOM _____ DATE _____

Lesson Plan for Block Scheduling

1-day lesson (See *Pacing the Chapter*, TE pages 318C–318D)   **For use with pages 347–355**

GOALS 1. **Use properties of sides and angles of rhombuses, rectangles, and squares.**
2. **Use properties of diagonals of rhombuses, rectangles, and squares.**

State/Local Objectives _____

_____

_____

✓ **Check the items you wish to use for this lesson.**

CHAPTER PACING GUIDE	
Day	**Lesson**
1	Assess Ch. 5; 6.1 (begin)
2	6.1 (end); 6.2 (begin)
3	6.2 (end); 6.3 (begin)
4	6.3 (end); **6.4 (begin)**
5	**6.4 (end)**; 6.5 (begin)
6	6.5 (end); 6.6 (all)
7	6.7 (all)
8	Review Ch. 6; Assess Ch. 6

STARTING OPTIONS
____ Homework Check: TE page 342: Answer Transparencies
____ Warm-Up or Daily Homework Quiz: TE pages 347 and
          345, CRB page 52, or Transparencies

TEACHING OPTIONS
____ Motivating the Lesson: TE page 348
____ Lesson Opener (Geometry Software): CRB page 53 or Transparencies
____ Technology Activity with Keystrokes: CRB pages 54–57
____ Examples: Day 4: 1–3, SE pages 347–348; Day 5: 4–6, SE pages 349–350
____ Extra Examples: Day 4: TE page 348 or Transp.; Day 5: TE pages 349–350 or Transp.; Internet
____ Closure Question: TE page 350
____ Guided Practice: SE page 351   Day 4: Exs. 1, 3–8, 10–11; Day 5: Exs. 2, 9

APPLY/HOMEWORK
Homework Assignment (See also the assignments for Lessons 6.3 and 6.5.)
____ Block Schedule: Day 4: 12–43; Day 5: 44–62, 66–68, 74–82 even, 83

Reteaching the Lesson
____ Practice Masters: CRB pages 58–60 (Level A, Level B, Level C)
____ Reteaching with Practice: CRB pages 61–62 or Practice Workbook with Examples
____ Personal Student Tutor

Extending the Lesson
____ Applications (Real-Life): CRB page 64
____ Challenge: SE page 355; CRB page 65 or Internet

ASSESSMENT OPTIONS
____ Checkpoint Exercises: Day 4: TE page 348 or Transp.; Day 5: TE pages 349–350 or Transp.
____ Daily Homework Quiz (6.4): TE page 355, CRB page 68, or Transparencies
____ Standardized Test Practice: SE page 355; TE page 355; STP Workbook; Transparencies

Notes _____

_____

_____

WARM-UP EXERCISES

For use before Lesson 6.4, pages 347–355

1. In $\square ABCD$, $m\angle A = (3x + 15)°$ and $m\angle C = (5x - 17)°$. What is the value of x?

2. Find the distance between $K(1, 3)$ and $M(3, 4)$.

3. In $\square KJLM$, $KJ = 10y - 5$ and $LM = -6y + 27$. What is the value of y?

4. The vertices of $PQRS$ are $P(-1, -3)$, $Q(2, -4)$, $R(5, -1)$, and $S(2, 0)$. Is $PQRS$ a parallelogram?

DAILY HOMEWORK QUIZ

For use after Lesson 6.3, pages 338–346

1. Describe how to prove that $ACEG$ is a parallelogram given that $\triangle BCD \cong \triangle FGH$ and $\triangle DEF \cong \triangle HAB$.

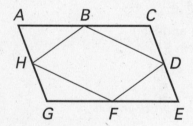

2. Prove that $EFGH$ is a parallelogram by showing that a pair of opposite sides are both congruent and parallel. Use $E(1, 2)$, $F(7, 9)$, $G(9, 8)$, and $H(3, 1)$.

3. Prove that $JKLM$ is a parallelogram by showing that the diagonals bisect each other. Use $J(-4, 4)$, $K(-1, 5)$, $L(1, -1)$, and $M(-2, -2)$.

Geometry Software Lesson Opener
For use with pages 347–355

Use geometry software to construct a parallelogram *ABFC* as described to the right. Display all side lengths and angle measures of *ABFC* on the screen.

To construct a parallelogram:

1. Draw a segment and label it $\overline{AB}$. From point *A*, draw another segment $\overline{AC}$.

2. Construct a line through *B* parallel to $\overline{AC}$.

3. Construct a line through *C* parallel to $\overline{AB}$.

4. Mark the intersection of these lines *F*. Hide the lines.

1. Drag $\overline{AB}$ until all four sides of *ABFC* are congruent. Now *ABFC* is a *rhombus*. Is it still a parallelogram?

2. Drag point *A* of rhombus *ABFC* until all four angles are congruent. (Make sure all four sides remain congruent.) Now *ABFC* is a *square*. Is it still a parallelogram? Is it still a rhombus? What is the measure of each angle of the square?

3. Drag $\overline{AB}$ of square *ABFC* until opposite sides are congruent but consecutive sides are not. (Make sure all four angles remain congruent.) Now *ABFC* is a *rectangle*. Is it still a parallelogram?

4. Drag point *A* of rectangle *ABFC* until it resembles your original parallelogram. By dragging point *A*, do you change *side lengths* or *angle measures* of the figure?

5. Continue to experiment with changing *ABFC* to various types of parallelograms. For each change in shape below, do you need to change *side lengths*, *angle measures*, or *both*?

   a. parallelogram to rectangle   **b.** rectangle to square

   c. square to rhombus   **d.** rhombus to parallelogram

   e. parallelogram to square   **f.** rectangle to rhombus

Technology Activity

For use with pages 347–355

GOAL **To use geometry software to verify statements about special parallelograms**

Geometry software can be used to verify statements about special parallelograms. For example, you could use geometry software to construct the rectangle below. Then, you could use the software's tools to verify the statement about diagonals of the rectangle.

Given: *ABCD* is a rectangle.

Prove: $\overline{AC} \cong \overline{BD}$

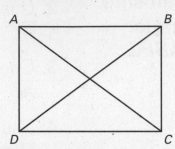

Activity

❶ Construct rectangle *ABCD* (see figure above.) Use the software's grid feature to ensure that you construct the sides such that the opposite sides are parallel and congruent.

❷ Construct the diagonals of the rectangle, $\overline{AC}$ and $\overline{BD}$.

❸ Measure the lengths of $\overline{AC}$ and $\overline{BD}$ and verify that $\overline{AC} \cong \overline{BD}$.

Exercises

Use geometry software to verify the following.

1. Given: *ABCD* is a parallelogram, $\overline{AC} \cong \overline{BD}$.
   Prove: *ABCD* is a rectangle.

2. Given: *ABCD* is a rhombus.
   Prove: $\overline{AC} \perp \overline{BD}$

3. Given: *ABCD* is a rhombus.
   Prove: $\overline{AC}$ bisects ∠*DAB* and ∠*BCD*
   and $\overline{BD}$ bisects ∠*ADC* and ∠*ABC*.

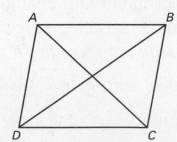

Lesson 6.4

NAME _____ DATE _____

Technology Activity Keystrokes

For use with pages 347–355

TI-92

1. Construct rectangle *ABCD*.

    [F8] 9 (Set Coordinate Axes to RECTANGULAR and Grid to ON.) [ENTER]

    [F3] 4 (Move cursor to point $(-2, 1)$ and prompt says, "ON THIS POINT OF
    THE GRID.") [ENTER] *A* (Move cursor to point $(2, 1)$ and prompt says,
    "ON THIS POINT OF THE GRID.") [ENTER] *B* (Move cursor to point $(2, -1)$
    and prompt says, "ON THIS POINT OF THE GRID.") [ENTER] *C* (Move
    cursor to point $(-2, 1)$ and prompt says, "ON THIS POINT OF THE GRID.")
    [ENTER] *D* (Move cursor to point *A*.) [ENTER]

2. Construct the diagonals of the rectangle, $\overline{AC}$ and $\overline{BD}$.

    [F2] 5 (Move cursor to point *A*.) [ENTER] (Move cursor to point *C*.) [ENTER]
    (Move cursor to point *B*.) [ENTER] (Move cursor to point *D*.) [ENTER]

3. Measure the lengths of $\overline{AC}$ and $\overline{BD}$.

    [F6] 1 (Move cursor to $\overline{AC}$.) [ENTER] (Move cursor to $\overline{BD}$.) [ENTER]

LESSON
6.4
CONTINUED

NAME _____ DATE _____

Technology Activity Keystrokes
For use with pages 347–355

SKETCHPAD

1. Turn on the axes and the grid. Choose **Snap To Grid** from the **Graph** menu.

   Choose the segment straightedge tool.

   Draw a segment from $(-2, 1)$ to $(2, 1)$.

   Draw a segment from $(2, 1)$ to $(2, -1)$.

   Draw a segment from $(2, -1)$ to $(-2, -1)$.

   Draw a segment from $(-2, -1)$ to $(-2, 1)$. Label the points.

   Choose the text tool. Label point $(-2, 1)$ A, label point $(2, 1)$ B, label point $(2, -1)$ C, and label point $(-2, -1)$ D.

   Turn off the axes and the grid.

   Choose **Hide Axes** from the **Graph** menu.

   Choose **Hide Grid** from the **Graph** menu.

2. Choose the segment straightedge tool and construct the diagonals of the rectangle, $\overline{AC}$ and $\overline{BD}$.

3. Measure the lengths of $\overline{AC}$ and $\overline{BD}$.

   Choose the selection arrow tool and select $\overline{AC}$. Then hold down the shift key, select $\overline{BD}$, and choose **Length** from the **Measure** menu.

Lesson 6.4

Technology Activity Keystrokes

For use with page 354

Keystrokes for Exercises 63–65

TI-92

1. Draw a segment $\overline{AB}$ and a point C on the segment.

   F2 5 (Place cursor at desired location for point A.) **ENTER** A (Move cursor to desired location for point B.) **ENTER** B

   F2 2 (Move cursor to location on segment.) **ENTER** C

2. Construct the midpoint of $\overline{AB}$ and label the point D.

   F4 3 (Move the cursor to $\overline{AB}$.) **ENTER** D

3. Hide $\overline{AB}$ and point B.

   F7 1 (Move cursor to $\overline{AB}$.) **ENTER** (Move cursor to point B.) **ENTER**

   (The hidden objects will remain dashed until the next command is given.)

4. Construct $\overline{AD}$.

   F2 5 (Place cursor on point A.) **ENTER** (Move cursor on point D.) **ENTER**

5. Construct two circles with centers A and C using length AD as the radius.

   F4 8 (Place cursor on point A.) **ENTER** (Place cursor on $\overline{AD}$.) **ENTER**

   F4 8 (Place cursor on point C.) **ENTER** (Place cursor on $\overline{AD}$.) **ENTER**

6. Hide $\overline{AD}$. (See Step 3.)

7. Label the intersection points of the two circles E and F.

   F2 3 (Move cursor to one intersection point.) **ENTER** E (Move cursor to other intersection point.) **ENTER** F

8. Draw $\overline{AE}$, $\overline{CE}$, $\overline{CF}$, and $\overline{AF}$.

   F2 5 (Place cursor on point A.) **ENTER** (Place cursor on point E.) **ENTER**
   Repeat for the other segments.

SKETCHPAD

1. Draw a segment $\overline{AB}$ and a point C on the segment. Choose segment from the straightedge tools and draw $\overline{AB}$. Choose the point tool and draw C on $\overline{AB}$.

2. Construct the midpoint of $\overline{AB}$ and label the point D. Use the selection arrow tool to select $\overline{AB}$. Choose **Point at Midpoint** from the **Construct** menu.

3. Hide $\overline{AB}$ and point B. Use the selection arrow tool to select $\overline{AB}$ and point B. Choose **Hide Objects** from the **Display** menu.

4. Construct $\overline{AD}$. Choose segment from the straightedge tools.

5. Construct two circles with centers A and C using length AD as the radius. Use the selection arrow tool to select $\overline{AD}$ and point A. Choose **Circle by Center and Radius** from the **Construct** menu.

6. Hide $\overline{AD}$. With the pointer tool select $\overline{AD}$. Choose **Hide Segment** from the **Display** menu.

7. Label the intersection points of the two circles E and F using the point tool.

8. Draw $\overline{AE}$, $\overline{CE}$, $\overline{CF}$, and $\overline{AF}$ using the segment straightedge tool.

Practice A

For use with pages 347–355

Each figure is a parallelogram. Identify the special type and explain your reasoning.

1.
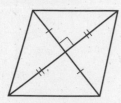

2.
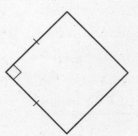

3.
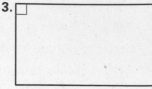

4.
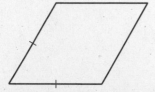

5.
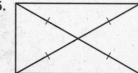

6.
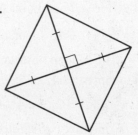

Match the properties of a quadrilateral with all of the types of quadrilateral which have that property.

7. The diagonals are congruent.

8. Both pairs of opposite sides are congruent.

9. Both pairs of opposite sides are parallel.

10. All angles are congruent.

11. All sides are congruent.

12. Diagonals bisect the angles.

A. Parallelogram

B. Rectangle

C. Rhombus

D. Square

MATH is a parallelogram with diagonals intersecting at O. Identify the type depending upon the given conditions.

13. $\overline{MT} \perp \overline{AH}$

14. $\overline{MT} \cong \overline{AH}$

15. $\overline{MA} \perp \overline{AT}, \overline{AM} \cong \overline{MH}$

16. $\overline{MO} \cong \overline{OT}, \overline{AO} \cong \overline{OH}$

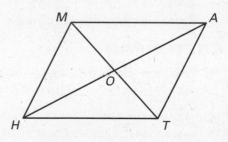

Find the value of x.

17. *MNOP* is a square.

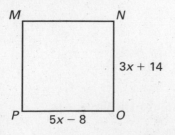

3x + 14

5x − 8

18. *DEFG* is a rhombus.

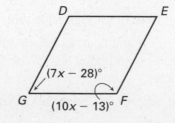

(7x − 28)°

(10x − 13)°

19. *WXYZ* is a rectangle.

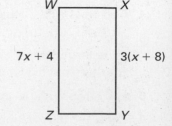

7x + 4

3(x + 8)

NAME _____ DATE _____

Reteaching with Practice

For use with pages 347–355

GOAL Use properties of sides and angles of rhombuses, rectangles, and squares and use properties of diagonals of rhombuses, rectangles, and squares

---

VOCABULARY

A **rhombus** is a parallelogram with four congruent sides.

A **rectangle** is a parallelogram with four right angles.

A **square** is a parallelogram with four congruent sides and four right angles.

Rhombus Corollary
A quadrilateral is a rhombus if and only if it has four congruent sides.

Rectangle Corollary
A quadrilateral is a rectangle if and only if it has four right angles.

Square Corollary
A quadrilateral is a square if and only if it is a rhombus and a rectangle.

Theorem 6.11
A parallelogram is a rhombus if and only if its diagonals are perpendicular.

Theorem 6.12
A parallelogram is a rhombus if and only if each diagonal bisects a pair of opposite angles.

Theorem 6.13
A parallelogram is a rectangle if and only if its diagonals are congruent.

EXAMPLE 1 *Using Properties of Special Parallelograms*

ABCD is a square. What else do you know
about *ABCD*?

SOLUTION

By the definition of a square, *ABCD* has four right angles and
four congruent sides. Also, because squares are parallelograms,
ABCD has all of the properties of a parallelogram:

- Opposite sides are parallel and congruent.

- Opposite angles are congruent and consecutive angles are
  supplementary.

- Diagonals bisect each other.

Exercises for Example 1

State all that you know about the special parallelogram given.

1. *ABCD* is a rhombus. What else do you know about *ABCD*?

2. *EFGH* is a rectangle. What else do you know about *EFGH*?

NAME _____ DATE _____

Reteaching with Practice

For use with pages 347–355

EXAMPLE 2 *Using Properties of Special Parallelograms*

For any rectangle *ABCD*, decide whether the statement is *always*, *sometimes*, or *never true*.

a. $\angle A \cong \angle C$ **b.** $\overline{AB} \cong \overline{CD}$ **c.** $\overline{AB} \cong \overline{BC}$

SOLUTION

a. Always true. By definition, a rectangle has four right angles, so $\angle A$ and $\angle C$ are both right angles, and therefore they are congruent.

b. Always true. $\overline{AB}$ and $\overline{CD}$ are opposite sides. Because all rectangles are parallelograms, opposite sides of a rectangle are congruent.

c. Sometimes true. $\overline{AB}$ and $\overline{BC}$ are adjacent sides. Adjacent sides of a rectangle do not need to be congruent.

Exercises for Example 2

For any rectangle *ABCD*, decide whether the statement is *always*, *sometimes*, or *never* true. Draw a sketch and explain your answer.

3. $\angle A \cong \angle C$ **4.** $\overline{AB} \cong \overline{BC}$ **5.** $\overline{AC} \cong \overline{BD}$

EXAMPLE 3 *Using Properties of a Rhombus*

In the diagram, *ABCD* is a rhombus.

What is the value of *x*?

SOLUTION

Because *ABCD* is a rhombus, all four sides are congruent. So, $AB = BC$.

$$4x - 5 = 2x + 17 \qquad \text{Equate lengths of congruent sides.}$$

$$x = 11 \qquad \text{Solve for } x.$$

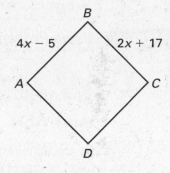

Exercises for Example 3

Use properties of special quadrilaterals to find the value of *x*.

6. *ABCD* is a rhombus.   **7.** *EFGH* is a rectangle.   **8.** *ABCD* is a square.

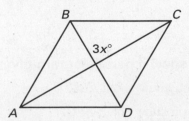   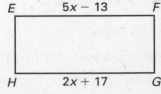   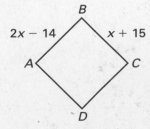

Lesson 6.4

NAME _____ DATE _____

Quick Catch-Up for Absent Students

For use with pages 347–355

The items checked below were covered in class on (date missed) _____

Lesson 6.4: Rhombuses, Rectangles, and Squares

____ **Goal 1**: Use properties of sides and angles of rhombuses, rectangles, and squares. (pp. 347–348)

Material Covered:

____ Example 1: Describing a Special Parallelogram

____ Example 2: Using Properties of Special Parallelograms

____ Student Help: Look Back

____ Example 3: Using Properties of a Rhombus

Vocabulary:

rhombus, p. 347   rectangle, p. 347
square, p. 347

____ **Goal 2**: Use properties of diagonals of rhombuses, rectangles, and squares. (pp. 349–350)

Material Covered:

____ Example 4: Proving Theorem 6.11

____ Example 5: Coordinate Proof of Theorem 6.11

____ Example 6: Checking a Rectangle

____ Other (specify) _____

_____

Homework and Additional Learning Support

____ Textbook (specify) _pp. 351–355_____

_____

____ Internet: Extra Examples at www.mcdougallittell.com

____ *Reteaching with Practice* worksheet (specify exercises)_____

____ *Personal Student Tutor* for Lesson 6.4

NAME _____ DATE _____

Real-Life Application: When Will I Ever Use This?

For use with pages 347–355

Quilting

Although quilting can be traced to prehistoric times, it flourished in Europe from the 17th through the 19th century. Quilting was brought over to America with the colonists, who often used the technique for such practical purposes as clothing and bedcovers. At first the designs imitated those of the English and Dutch, but soon an American style quilt took form.

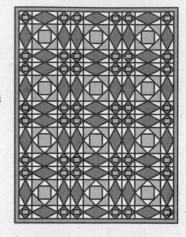

Using such techniques as patchwork and applique, quilts commonly display geometric patterns or pictures of animals, people, and objects in nature. Often quilts are used to remember historical events, important family memories, or stories. They have truly become a symbolic representation of America's heritage.

More traditional quilts are fashioned by piecing together quilt blocks. In the illustration of the Storm at Sea quilt block below, you can see how the repetition of the pattern creates the finished product at the right.

In Exercises 1 and 2, refer to the quilt block below.

1. Use as many words as possible to describe each shape from the quilt block: *rhombus*, *rectangle*, *square*. You may wish to use a centimeter ruler and/or a protractor.

a.   b.   c.

2. Copy and complete the table by counting the number of each shape.

Shape	Rhombus	Rectangle	Square
Number	_____	_____	_____

Lesson 6.4

NAME _____ DATE _____

Challenge: Skills and Applications

For use with pages 347–355

In Exercises 1–5, find all possible values of *x* and *y*. Diagrams may not be drawn to scale.

1. *ABCD* is a rectangle.

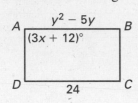

2. *EFGH* is a rhombus.

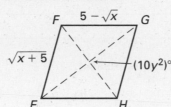

3. *IJKL* is a rectangle.

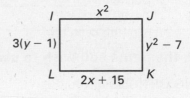

4. *MNOP* is a rectangle.

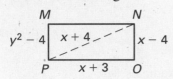

5. *UVWX* is a parallelogram but not a rhombus.

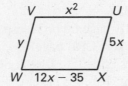

6. Refer to the diagram. Write a two-column proof.

Given: $\overline{BD}$ is the perpendicular bisector of $\overline{AC}$, $\overline{AD} \parallel \overline{BC}$.

Prove: *ABCD* is a rhombus.

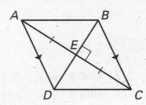

7. Refer to the diagram. Write a paragraph proof.

Given: *IJKL* is a parallelogram, $\angle 1 \cong \angle 2$.

Prove: *IJKL* is a rectangle.

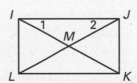

8. Refer to the diagram. Write a two-column proof.

Given: *WXYZ* is a parallelogram, $\angle 1 \cong \angle 2$.

Prove: *WXYZ* is a rhombus.

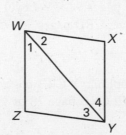

9. A construction worker wants to make sure that an opening for a window is rectangular. If both diagonals measure exactly 37 inches, can the worker be certain that the opening is rectangular? Explain.

TEACHER'S NAME _____ CLASS _____ ROOM _____ DATE _____

Lesson Plan

2-day lesson (See *Pacing the Chapter,* TE pages 318C–318D)   **For use with pages 356–363**

GOALS   1. **Use properties of trapezoids.**
            2. **Use properties of kites.**

State/Local Objectives _____

_____

✓ Check the items you wish to use for this lesson.

STARTING OPTIONS
____ Homework Check: TE page 351: Answer Transparencies
____ Warm-Up or Daily Homework Quiz: TE pages 356 and 355, CRB page 68, or Transparencies

TEACHING OPTIONS
____ Motivating the Lesson: TE page 357
____ Lesson Opener (Application): CRB page 69 or Transparencies
____ Technology Activity with Keystrokes: CRB page 70
____ Examples: Day 1: 1–5, SE pages 356–358; Day 2: See the Extra Examples.
____ Extra Examples: Day 1 or Day 2: 1–5, TE pages 357–358 or Transp.; Internet
____ Closure Question: TE page 358
____ Guided Practice: SE page 359   Day 1: Exs. 1–9; Day 2: See Checkpoint Exs. TE pages 357–358

APPLY/HOMEWORK
Homework Assignment
____ Basic   Day 1: 10–24 even, 25–27, 28–40 even, 46, 48, 50–52; Day 2: 11–23 odd, 29–39 odd,
             47, 49, 54–64 even; Quiz 2: 1–6
____ Average   Day 1: 10–24 even, 25–27, 28–40 even, 46, 48, 50–52; Day 2: 11–23 odd,
             29–41 odd, 47, 49, 54–64 even; Quiz 2: 1–6
____ Advanced   Day 1: 10–24 even, 25–27, 28–40 even, 46, 48, 50–53; Day 2: 11–23 odd,
             29–41 odd, 47, 49, 54–64 even; Quiz 2: 1–6

Reteaching the Lesson
____ Practice Masters: CRB pages 71–73 (Level A, Level B, Level C)
____ Reteaching with Practice: CRB pages 74–75 or Practice Workbook with Examples
____ Personal Student Tutor

Extending the Lesson
____ Applications (Interdisciplinary): CRB page 77
____ Challenge: SE page 362; CRB page 78 or Internet

ASSESSMENT OPTIONS
____ Checkpoint Exercises: Day 1 or Day 2: TE pages 357–358 or Transp.
____ Daily Homework Quiz (6.5): TE page 363, CRB page 82, or Transparencies
____ Standardized Test Practice: SE page 362; TE page 363; STP Workbook; Transparencies
____ Quiz (6.4–6.5): SE page 363; CRB page 79

Notes _____

_____

_____

TEACHER'S NAME _____ CLASS _____ ROOM _____ DATE _____

Lesson Plan for Block Scheduling

1-day lesson (See *Pacing the Chapter,* TE pages 318C–318D)   For use with pages 356–363

 GOALS   1. **Use properties of trapezoids.**
                2. **Use properties of kites.**

State/Local Objectives _____

_____

_____

CHAPTER PACING GUIDE	
Day	**Lesson**
1	Assess Ch. 5; 6.1 (begin)
2	6.1 (end); 6.2 (begin)
3	6.2 (end); 6.3 (begin)
4	6.3 (end); 6.4 (begin)
5	6.4 (end); **6.5 (begin)**
6	**6.5 (end)**; 6.6 (all)
7	6.7 (all)
8	Review Ch. 6; Assess Ch. 6

✓ **Check the items you wish to use for this lesson.**

STARTING OPTIONS

____ Homework Check: TE page 351: Answer Transparencies
____ Warm-Up or Daily Homework Quiz: TE pages 356 and
         355, CRB page 68, or Transparencies

TEACHING OPTIONS

____ Motivating the Lesson: TE page 357
____ Lesson Opener (Application): CRB page 69 or Transparencies
____ Technology Activity with Keystrokes: CRB page 70
____ Examples: Day 5: 1–5, SE pages 356–358; Day 6: See the Extra Examples.
____ Extra Examples: Day 5 or Day 6: 1–5, TE pages 357–358 or Transp.; Internet
____ Closure Question: TE page 358
____ Guided Practice: SE page 359   Day 5: Exs. 1–9; Day 6: See Checkpoint Exs. TE pages 357–358

APPLY/HOMEWORK

Homework Assignment (See also the assignments for Lessons 6.4 and 6.6.)
____ Block Schedule: Day 5: 10–24 even, 25–27, 28–42 even, 46, 48, 50–52; Day 6: 11–23 odd,
         29–41 odd, 47, 49, 54–64 even; Quiz 2: 1–6

Reteaching the Lesson
____ Practice Masters: CRB pages 71–73 (Level A, Level B, Level C)
____ Reteaching with Practice: CRB pages 74–75 or Practice Workbook with Examples
____ Personal Student Tutor

Extending the Lesson
____ Applications (Interdisciplinary): CRB page 77
____ Challenge: SE page 362; CRB page 78 or Internet

ASSESSMENT OPTIONS

____ Checkpoint Exercises: Day 5 or Day 6: TE pages 357–358 or Transp.
____ Daily Homework Quiz (6.5): TE page 363, CRB page 82, or Transparencies
____ Standardized Test Practice: SE page 362; TE page 363; STP Workbook; Transparencies
____ Quiz (6.4–6.5): SE page 363; CRB page 79

Notes _____

_____

_____

LESSON 6.5

Available as a transparency

NAME _____ DATE _____

WARM-UP EXERCISES

For use before Lesson 6.5, pages 356–363

Find the slope of the line passing through the given points.

1. $(3, 5)$ and $(7, 1)$ **2.** $(-2, -2)$ and $(0, 0)$

Triangle *ABC* is a right triangle with hypotenuse $\overline{AB}$. Use the Pythagorean theorem to find each length.

3. AB, given that $AC = 8$ and $BC = 10$

4. BC, given that $AC = 5$ and $AB = 13$

DAILY HOMEWORK QUIZ

For use after Lesson 6.4, pages 347–355

For any rhombus *ABCD*, decide whether the statement in Exercises 1–3 is *always*, *sometimes*, or *never true*. If the answer is not *always*, explain.

1. $\overline{AC} \perp \overline{BD}$ **2.** $\overline{AC} \cong \overline{BD}$ **3.** $\overline{AB} \parallel \overline{CD}$

4. For which parallelogram is it true both that the diagonals are congruent and that each diagonal bisects a pair of opposite angles?

5. Find the value of *x* in rectangle *PQRS* if $PT = 5x + 1$ and $QT = 3x + 3$.

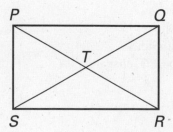

Geometry
Chapter 6 Resource Book

NAME _____ DATE _____

Application Lesson Opener

For use with pages 356–363

1. The shape of the driver's side window on the vehicle below is a *trapezoid*. A **trapezoid** is a quadrilateral with exactly one pair of parallel sides. Sketch the window and mark the parallel sides with arrows. Sketch the other two side windows, mark any parallel sides, and name the shapes.

2. The front and back windows of the vehicle are *isosceles* trapezoids. Sketch one of these windows and mark the parallel sides with arrows. What do you think makes a trapezoid *isosceles*? What might be some properties of an isosceles trapezoid?

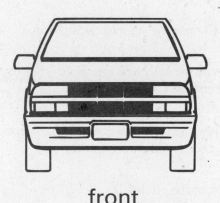 

front   back

NAME _____ DATE _____

Technology Activity Keystrokes

For use with page 361

Keystrokes for Exercises 43–45

TI-92

1. Draw points *A*, *B*, and *C*.

   `F2` 1 `ENTER` *A* (Move cursor to location for point *B*.) `ENTER` *B*

   (Move cursor to location of point *C*.) `ENTER` *C*

   Draw segments $\overline{AC}$ and $\overline{BC}$.

   (Move cursor to point *A*.) `F2` 5 `ENTER` (Move cursor to point *C*.) `ENTER`

   `ENTER` (Move cursor to point *B*.) `ENTER`

2. Construct a circle with center *A* and radius *AC*.

   (Move cursor to point *A*.) `F3` 1 `ENTER` (Move cursor to point *C*.) `ENTER`

3. Construct a circle with center *B* and radius *BC*.

   (Move cursor to point *B*.) `F3` 1 `ENTER` (Move cursor to point *C*.) `ENTER`

4. Label the other intersection of the circles *D*.

   (Move to intersection of the circles that is not labeled *C*.) `F2` 3 `ENTER` *D*

5. Draw $\overline{BD}$ and $\overline{AD}$.

   (Move cursor to point *D*.) `F2` 5 `ENTER` (Move cursor to point *B*.) `ENTER`

   (Move cursor to point *D*.) `ENTER` (Move cursor to point *A*.) `ENTER`

SKETCHPAD

1. Choose the point tool and draw points *A*, *B*, and *C*. Choose the segment straightedge tool and draw segments $\overline{AC}$ and $\overline{BC}$.

2. Construct a circle with center *A* and radius *AC*. Choose the selection arrow tool. Select point *A*, hold down the shift key and select $\overline{AC}$, and choose **Circle by Center and Radius** from the **Construct** menu.

3. Construct a circle with center *B* and radius *BC*. Choose the selection arrow tool. Select point *B*, hold down the shift key and select $\overline{BC}$, and choose **Circle by Center and Radius** from the **Construct** menu.

4. Choose the point tool and label the other intersection of the circles *D*.

5. Choose the segment straightedge tool and draw segments $\overline{BD}$ and $\overline{AD}$.

NAME _____ DATE _____

Practice A

For use with pages 356–363

Match the pair of segments or angles with the term that describes them in trapezoid PQRS.

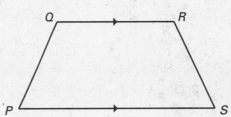

1. $\overline{QR}$ and $\overline{PS}$ **A.** bases

2. $\overline{PQ}$ and $\overline{RS}$ **B.** legs

3. $\overline{QS}$ and $\overline{PR}$ **C.** diagonals

4. $\angle Q$ and $\angle S$ **D.** base angles

5. $\angle S$ and $\angle P$ **E.** opposite angles

Complete the statement with always, sometimes or never.

6. A trapezoid is __?__ a parallelogram.

7. The bases of a trapezoid are __?__ parallel.

8. The base angles of an isosceles trapezoid are __?__ congruent.

9. The legs of a trapezoid are __?__ congruent.

Find the angle measures of ABCD.

10.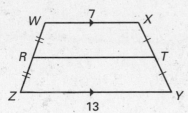

11.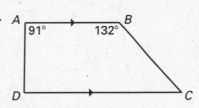

12.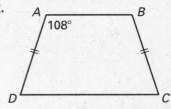

Find the length of the midsegment $\overline{RT}$.

13.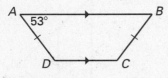

14.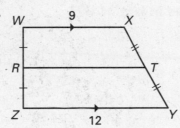

15.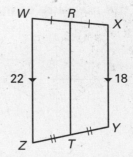

Find the length of the sides to the nearest hundredth or the measure of the angles in kite KITE.

16.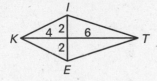

17.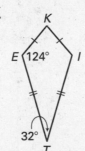

18.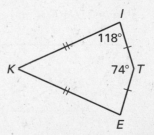

NAME _____   DATE _____

Practice B

For use with pages 356–363

Draw a trapezoid *JKLM* with $\overline{JK} \parallel \overline{LM}$. Match the pair of segments or angles with the term that describes them in trapezoid *JKLM*.

1. $\overline{JK}$ and $\overline{ML}$ **2.** $\overline{MJ}$ and $\overline{KL}$ **3.** $\overline{ML}$ and $\overline{KL}$

4. $\angle K$ and $\angle M$ **5.** $\overline{JL}$ and $\overline{KM}$ **6.** $\angle M$ and $\angle L$

   A. bases angles   **B.** consecutive sides   **C.** opposite angles

   D. diagonals   **E.** bases   **F.** legs

Find the angle measures of *ABCD*.

7.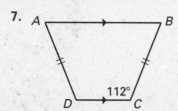

8.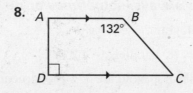

9.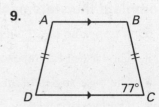

The midsegment of the trapezoid is $\overline{RT}$. Find the value of *x*.

10.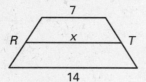

11.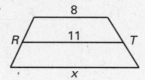

12.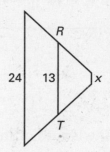

Find the length of the sides to the nearest hundredth or the measure of the angles in kite *MATH*.

13.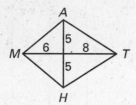

14.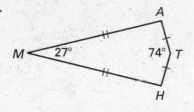

15.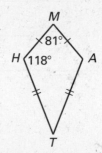

Write a two-column or a paragraph proof.

16. Given: $\overline{DE} \parallel \overline{AV}$,

       $\triangle DAV \cong \triangle EVA$

Prove: *DAVE* is an isosceles trapezoid.

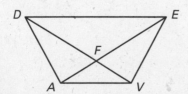

17 Given: $\overline{WV}$ is a midsegment of $\triangle XYZ$.

       $\overline{XZ} \cong \overline{YZ}$

Prove: *XWVY* is an isosceles trapezoid.

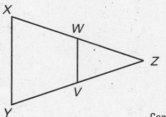

NAME _____   DATE _____

Practice C
For use with pages 356–363

Decide whether the figure is a trapezoid. If it is, is it an isosceles trapezoid?

1.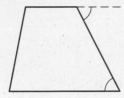

2.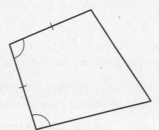

3.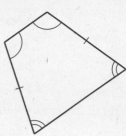

Quadrilateral *ABCD* is a trapezoid with midsegment $\overline{EF}$. Use the given information to answer the following.

4. If $m\angle B = 73°$, then $m\angle C =$ ___?___ .

5. If $m\angle A = 51°$ and $m\angle C = 105°$, then $m\angle D =$ ___?___ .

6. If $m\angle A = 48°$ and $m\angle C = 112°$, then $m\angle CFE =$ ___?___ .

7. If $AB = 28$ and $DC = 13$, then $EF =$ ___?___ .

8. If $EF = 13$ and $DC = 6$, then $AB =$ ___?___ .

9. If $EF = x + 5$ and $DC + AB = 4x + 6$, then $EF =$ ___?___ .

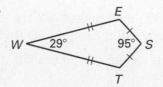

Find the length of the sides to the nearest hundredth, or the measure of the angles in kite *WEST*.

10.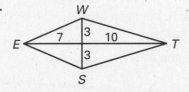

11.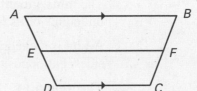

12.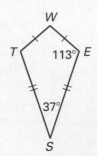

13. In an isosceles trapezoid, if one pair of base angles is twice the measure of the second pair of base angles, what are the measures of the angles?

14. If the midsegment of a trapezoid measures 6 units long, what is true about the lengths of the bases of the trapezoid?

Write a two-column or a paragraph proof.

15. **Given:** *LORI* is a rectangle.

    $\overline{LB} \cong \overline{DO}$

    Prove: *BIRD* is an isosceles trapezoid.

    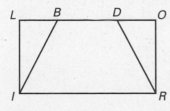

16. **Given:** $\overline{AF} \not\cong \overline{BC}$

    $\triangle ABC \cong \triangle CDA$

    Prove: *ABCF* is a trapezoid.

    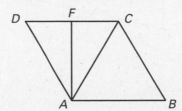

LESSON

6.5

Reteaching with Practice

For use with pages 356–363

GOAL Use properties of trapezoids and kites

VOCABULARY

A **trapezoid** is a quadrilateral with exactly one pair of parallel sides. The parallel sides of a trapezoid are the **bases** of the trapezoid.

For each of the bases of a trapezoid, there is a pair of **base angles,** which are the two angles that have that base as a side.

The nonparallel sides of a trapezoid are the **legs** of the trapezoid.

If the legs of a trapezoid are congruent, then the trapezoid is an **isosceles trapezoid.**

The **midsegment** of a trapezoid is the segment that connects the midpoints of its legs.

A **kite** is a quadrilateral that has two pairs of consecutive congruent sides, but opposite sides are not congruent.

Theorem 6.14 If a trapezoid is isosceles, then each pair of base angles is congruent.

Theorem 6.15 If a trapezoid has a pair of congruent base angles, then it is an isosceles trapezoid.

Theorem 6.16 A trapezoid is isosceles if and only if its diagonals are congruent.

Theorem 6.17 The midsegment of a trapezoid is parallel to each base and its length is one half the sum of the lengths of its bases.

Theorem 6.18 If a quadrilateral is a kite, then its diagonals are perpendicular.

Theorem 6.19 If a quadrilateral is a kite, then exactly one pair of opposite angles is congruent.

EXAMPLE 1 *Finding Midsegment Lengths of Trapezoids and Using Algebra*

---

a. Find the length of the midsegment $\overline{MN}$.

b. Find the value of x.

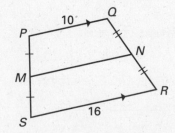

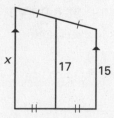

Reteaching with Practice

For use with pages 356–363

SOLUTION

a. Use the Midsegment Theorem for Trapezoids.

$$MN = \tfrac{1}{2}(PQ + SR) = \tfrac{1}{2}(10 + 16) = \tfrac{1}{2}(26) = 13$$

b. $17 = \tfrac{1}{2}(15 + x)$ Midsegment Theorem for Trapezoids

    $34 = 15 + x$ Multiply each side by 2.

    $19 = x$ Subtract.

Exercises for Example 1

Find the value of x.

1.

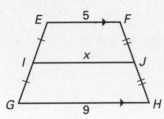

2.

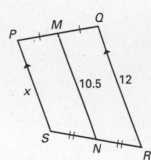

3.

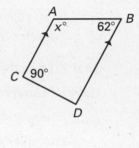

EXAMPLE 2   *Using Properties of Kites*

$JKLM$ is a kite. What is $m\angle J$?

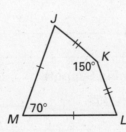

SOLUTION

$JKLM$ is a kite, so $\angle J \cong \angle L$ and $m\angle J = m\angle L$.

    $2(m\angle J) + 150° + 70° = 360°$ Sum of measures of int. ⊿s of a quad. is 360°.

           $2(m\angle J) = 140°$ Simplify.

              $m\angle J = 70°$ Divide each side by 2.

Exercises for Example 2

Find the value of x.

4.

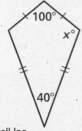

5.

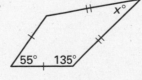

6.

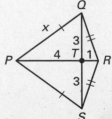

NAME _____ DATE _____

Quick Catch-Up for Absent Students

For use with pages 356–363

The items checked below were covered in class on (date missed) _____

Lesson 6.5: Trapezoids and Kites

_____ **Goal 1**: Use properties of trapezoids. (pp. 356–357)

Material Covered:

_____ Example 1: Using Properties of Isosceles Trapezoids

_____ Example 2: Using Properties of Trapezoids

_____ Example 3: Finding Midsegment Lengths of Trapezoids

Vocabulary:

    trapezoid, p. 356   bases, p. 356

    base angles, p. 356   legs, p. 356

    isosceles trapezoid, p. 356   midsegment, p. 357

_____ **Goal 2**: Use properties of kites. (p. 358)

Material Covered:

_____ Example 4: Using the Diagonals of a Kite

_____ Example 5: Angles of a Kite

Vocabulary:

    kite, p. 358

_____ Other (specify) _____

_____

Homework and Additional Learning Support

_____ Textbook (specify) _pp. 359–363_____

_____

_____ Internet: Extra Examples at www.mcdougallittell.com

_____ *Reteaching with Practice* worksheet (specify exercises)_____

_____ *Personal Student Tutor* for Lesson 6.5

NAME _____ DATE _____

Interdisciplinary Application

For use with pages 356–363

Spider Webs

LIFE SCIENCE   Although there are spiders that jump or chase their prey, the spider's web is the most well known way for spiders to catch their food. Using abdominal glands that produce a liquid silk, a spider secretes the silk through *spinnerets* to weave an elaborate snare for flying insects. With its highly developed sense of touch, the spider senses the vibrations of a caught insect in the web and then quickly gets its victim.

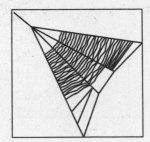
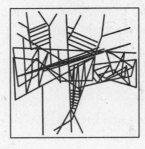
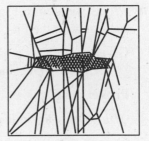

Not all spiders spin the same type of web. Four common varieties, the orb web, the triangle web, the tangle web, and the sheet web are shown at the right. Depending on the species and the design, a web can be completed in as little as 40 minutes to as long as several hours.

The orb web is probably the design that first comes to mind. In Exercises 1–4, use the diagram of an orb web shown below.

1. Decide whether the quadrilateral is a *trapezoid*, an *isosceles trapezoid*, a *kite*, or *none of these*. Explain your reasoning.

   a. *ABDC*

   b. *GHFD*

2. *ABFE* is a trapezoid. Prove *ABFE* is an isosceles trapezoid.

3. Find the length of the midsegment $\overline{CD}$.

4. Find the length of $\overline{JB}$.

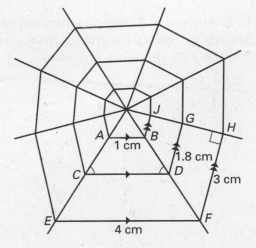

NAME _____ DATE _____

Challenge: Skills and Applications

For use with pages 356–363

In Exercises 1–3, each figure shown is a trapezoid with its midsegment. Find all possible values of _x_. Diagrams may not be drawn to scale.

1.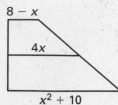
$8 - x$
$4x$
$x^2 + 10$

2.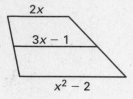
$2x$
$3x - 1$
$x^2 - 2$

3.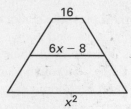
16
$6x - 8$
x^2

In Exercises 4–5, write a paragraph proof.

4. Prove Theorem 6.15: If a trapezoid has a pair of congruent base angles, then it is an isosceles trapezoid.

   Given: $ABCD$ is a trapezoid,
   $\overline{AB} \parallel \overline{DC}$, and $\angle D \cong \angle C$.

   Prove: $ABCD$ is an isosceles trapezoid.

   (*Hint:* Draw an additional line segment.)

5. **Given:** $\overline{PR} \perp \overline{QS}$, $\overline{PQ} \cong \overline{PS}$,

   T is not the midpoint of $\overline{PR}$.

   Prove: $PQRS$ is a kite.

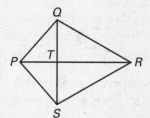

In Exercises 6–8, the coordinates of three vertices of an isosceles trapezoid are given. What are the coordinates of the remaining vertex? (Find or describe all possible answers.)

6. $(0, 0)$, $(2, 2)$, $(4, 2)$

7. $(0, 0)$, $(26, 0)$, $(0, 39)$

8. $(-5, 10)$, $(0, 0)$, $(5, 10)$

LESSON 6.5

Quiz 2

For use after Lessons 6.4 and 6.5

Determine whether *ABCD* is a *rectangle*, a *rhombus*, a *square*, a *trapezoid*, or a *kite*. (Lessons 6.4 and 6.5)

1. $A(5, 2), B(1, 9), C(-3, 2), D(1, -5)$

2. $A(2, 3), B(2, 6), C(-2, 6), D(-2, 3)$

3. $A(-5, -2), B(1, -2), C(3, 0), D(1, 4)$

4. $A(4, 0), B(0, 2), C(-7, 0), D(0, -2)$

5. In the figure at the right, $\overline{RS} \parallel \overline{TQ} \parallel \overline{UV}$ and $\overline{XY} \parallel \overline{QV}$. Find the total number of trapezoids in the figure. *(Lesson 6.5)*

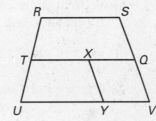

Answers
1. _____
2. _____
3. _____
4. _____
5. _____
6. _____

6. Write a two-column proof. *(Lesson 6.5)*
   Given: *ABCD* is an isosceles trapezoid.
   $\overline{DE} \perp \overline{AB}$; $\overline{CF} \perp \overline{AB}$
   Prove: $\overline{AD} \cong \overline{BC}$

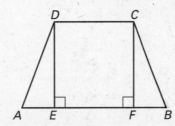

Lesson Plan

1-day lesson (See *Pacing the Chapter*, TE pages 318C–318D)   For use with pages 364–370

GOALS 1. Identify special quadrilaterals based on limited information.
2. Prove that a quadrilateral is a special type of quadrilateral, such as a rhombus or a trapezoid.

State/Local Objectives _____

_____

✓ **Check the items you wish to use for this lesson.**

STARTING OPTIONS
_____ Homework Check: TE page 359: Answer Transparencies
_____ Warm-Up or Daily Homework Quiz: TE pages 364 and 363, CRB page 82, or Transparencies

TEACHING OPTIONS
_____ Motivating the Lesson: TE page 365
_____ Lesson Opener (Visual Approach): CRB page 83 or Transparencies
_____ Examples 1–5: SE pages 364–366
_____ Extra Examples: TE pages 365–366 or Transparencies
_____ Closure Question: TE page 366
_____ Guided Practice Exercises: SE page 367

APPLY/HOMEWORK
Homework Assignment
_____ Basic   8–36, 38, 40, 45–47, 49, 53, 56–64 even
_____ Average   8–36, 38, 40, 45–50, 53, 56–64 even
_____ Advanced   8–40, 42–54, 56–64 even

Reteaching the Lesson
_____ Practice Masters: CRB pages 84–86 (Level A, Level B, Level C)
_____ Reteaching with Practice: CRB pages 87–88 or Practice Workbook with Examples
_____ Personal Student Tutor

Extending the Lesson
_____ Cooperative Learning Activity: CRB page 90
_____ Applications (Real-Life): CRB page 91
_____ Challenge: SE page 370; CRB page 92 or Internet

ASSESSMENT OPTIONS
_____ Checkpoint Exercises: TE pages 365–366 or Transparencies
_____ Daily Homework Quiz (6.6): TE page 370, CRB page 95, or Transparencies
_____ Standardized Test Practice: SE page 370; TE page 370; STP Workbook; Transparencies

Notes _____

_____

_____

TEACHER'S NAME _____ CLASS _____ ROOM _____ DATE _____

Lesson Plan for Block Scheduling

Half-day lesson (See *Pacing the Chapter*, TE pages 318C–318D)   For use with pages 364–370

GOALS 1. **Identify special quadrilaterals based on limited information.**
2. **Prove that a quadrilateral is a special type of quadrilateral, such as a rhombus or a trapezoid.**

State/Local Objectives _____

_____

_____

✓ **Check the items you wish to use for this lesson.**

STARTING OPTIONS

____ Homework Check: TE page 359; Answer Transparencies
____ Warm-Up or Daily Homework Quiz: TE pages 364 and
        363, CRB page 82, or Transparencies

TEACHING OPTIONS

____ Motivating the Lesson: TE page 365
____ Lesson Opener (Visual Approach): CRB page 83 or Transparencies
____ Examples 1–5: SE pages 364–366
____ Extra Examples: TE pages 365–366 or Transparencies
____ Closure Question: TE page 366
____ Guided Practice Exercises: SE page 367

APPLY/HOMEWORK

Homework Assignment (See also the assignment for Lesson 6.5.)
____ Block Schedule: 8–36, 38, 40, 45–50, 53, 56–64 even

Reteaching the Lesson

____ Practice Masters: CRB pages 84–86 (Level A, Level B, Level C)
____ Reteaching with Practice: CRB pages 87–88 or Practice Workbook with Examples
____ Personal Student Tutor

Extending the Lesson

____ Cooperative Learning Activity: CRB page 90
____ Applications (Real-Life): CRB page 91
____ Challenge: SE page 370; CRB page 92 or Internet

ASSESSMENT OPTIONS

____ Checkpoint Exercises: TE pages 365–366 or Transparencies
____ Daily Homework Quiz (6.6): TE page 370, CRB page 95, or Transparencies
____ Standardized Test Practice: SE page 370; TE page 370; STP Workbook; Transparencies

CHAPTER PACING GUIDE	
Day	Lesson
1	Assess Ch. 5; 6.1 (begin)
2	6.1 (end); 6.2 (begin)
3	6.2 (end); 6.3 (begin)
4	6.3 (end); 6.4 (begin)
5	6.4 (end); 6.5 (begin)
6	6.5 (end); **6.6 (all)**
7	6.7 (all)
8	Review Ch. 6; Assess Ch. 6

Lesson 6.6

Notes _____

_____

_____

NAME _____ DATE _____

WARM-UP EXERCISES

For use before Lesson 6.6, pages 364–370

Name the figure.

1. a quadrilateral with exactly one pair of opposite angles congruent and perpendicular diagonals

2. a quadrilateral that is both a rhombus and a rectangle

3. a quadrilateral with exactly one pair of parallel sides

4. any parallelogram with perpendicular diagonals

··

DAILY HOMEWORK QUIZ

For use after Lesson 6.5, pages 356–363

1. In isosceles trapezoid $EFGH$, $m\angle E = 106°$, and $\overline{EH}$ is a base. What are the measures of the other angles?

2. A nonisosceles trapezoid has one base of length 37 and a midsegment of length 29. What is the length of the second base?

3. To the nearest hundredth, what are the lengths of the sides of the kite?

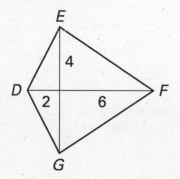

4. Determine whether the points $A(-3, 1)$, $B(-1, 5)$, $C(3, 7)$, and $D(5, -2)$ are the vertices of a kite. Explain.

Geometry
Chapter 6 Resource Book

NAME _____ DATE _____

Visual Approach Lesson Opener

For use with pages 364–370

Which figure do you think does not belong in a set with the other three? Explain why it does not belong. There may be more than one possible answer.

1.

rhombus   square   trapezoid   parallelogram

2.

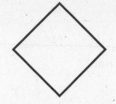 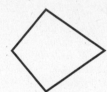

isosceles
trapezoid   kite   rhombus   square

3.

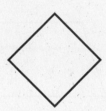 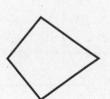 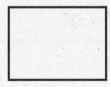

rhombus   trapezoid   kite   rectangle

4.

parallelogram   trapezoid   rectangle   isosceles
trapezoid

NAME _____ DATE _____

Practice A

For use with pages 364–370

Match the property on the left with all of the quadrilaterals that have the property.

1. Both pairs of opposite sides are parallel.
2. Both pairs of opposite sides are congruent.
3. Both pairs of opposite angles are congruent.
4. Exactly one pair of opposite sides are parallel.
5. Exactly one pair of opposite sides are congruent.
6. Exactly one pair of opposite angles are congruent.
7. Diagonals are congruent.
8. Diagonals are perpendicular.

   A. Parallelogram
   B. Rectangle
   C. Rhombus
   D. Square
   E. Trapezoid
   F. Isosceles Trapezoid
   G. Kite

Identify the special quadrilateral. Use the most specific name.

9.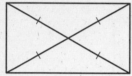

10.

11.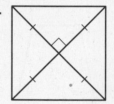

Which two segments or angles must be congruent to enable you to prove *ABCD* is the given quadrilateral? Explain your reasoning. There may be more than one right answer.

12. rectangle

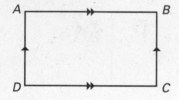

13. parallelogram

14. isosceles trapezoid

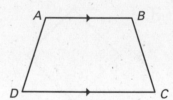

What kind of quadrilateral is *PQRS*? Justify your answer.

15.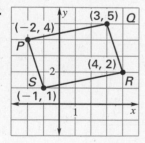

16.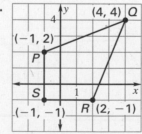

Copy the chart. Put an X in the box if the shape *always* has the given property.

Property	▱	Rectangle	Rhombus	Square	Trapezoid	Kite
1. Both pairs of opposite sides are congruent.						
2. Diagonals are congruent.						
3. Diagonals are perpendicular.						
4. Diagonals bisect one another.						
5. Consecutive angles are supplementary.						
6. Both pairs of opposite angles are congruent.						

What quadrilaterals meet the conditions shown? *ABCD* is not drawn to scale.

7.
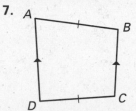

8.
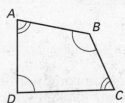

9.
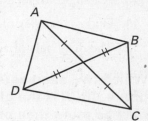

Which two segments or angles must be congruent to enable you to prove *ABCD* is the given quadrilateral? Explain your reasoning. There may be more than one right answer.

10. rhombus
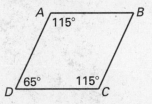

11. isosceles trapezoid
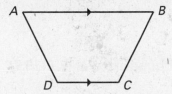

12. square
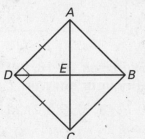

In Exercises 13–15, what kind of quadrilateral is *PQRS*? Justify your answer.

13. $P(5, 4)$, $Q(3, -6)$, $R(0, -10)$, $S(2, 0)$

14. $P(4, 8)$, $Q(0, 9)$, $R(-2, 1)$, $S(2, 0)$

15. $P(1, 5)$, $Q(8, 6)$, $R(15, 5)$, $S(8, 4)$

16. Use the quadrilateral in Exercise 15. Find the midpoint of each side. Connect the midpoints to form a new quadrilateral. What kind of quadrilateral is formed?

Lesson 6.6

Lesson 6.6

Draw the sides or diagonals of *ABCD* as described. What special type of quadrilateral is *ABCD*?

1. $\overline{AC} \cong \overline{BD}$, $\overline{AC}$ and $\overline{BD}$ bisect one another, but $\overline{AC}$ is not perpendicular to $\overline{BD}$.

2. $\overline{AB} \cong \overline{BC}$ and $\overline{CD} \cong \overline{DA}$, but $\overline{BC} \not\cong \overline{CD}$.

3. $\overline{AB} \parallel \overline{CD}$ and $\overline{BC} \cong \overline{DA}$.

4. $\overline{AC} \perp \overline{BD}$, $\overline{AC}$ and $\overline{BD}$ bisect one another, but $\overline{AC} \not\cong \overline{BD}$.

5. $\overline{AC} \perp \overline{BD}$, $\overline{AC}$ and $\overline{BD}$ bisect one another, and $\overline{AC} \cong \overline{BD}$.

Determine whether the statement is *always*, *sometimes*, or *never* true.

6. Diagonals of a trapezoid are congruent.

7. Opposite sides of a rectangle are congruent.

8. A square is a rectangle.

9. A square is not a rhombus.

10. All angles of a parallelogram are congruent.

11. Opposite angles of an isosceles trapezoid are congruent.

12. The diagonals of a parallelogram are perpendicular.

Which two segments or angles must be congruent to enable you to prove *ABCD* is the given quadrilateral? Explain your reasoning. There may be more than one right answer.

13. rectangle

14. kite

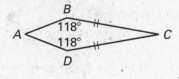

15. isosceles trapezoid

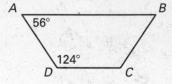

In Exercises 16–18, what kind of quadrilateral is *PQRS*? Justify your answer.

16. $P(-1, 3)$, $Q(4, 2)$, $R(1, -1)$, $S(-4, 0)$

17. $P(-3, 5)$, $Q(-7, 6)$, $R(-9, -2)$, $S(-5, -3)$

18. $P(-2, 9)$, $Q(-2, -1)$, $R(-5, 5)$, $S(-5, 7)$

19. Use the quadrilateral in Exercise 17. Find the midpoint of each side. Connect the midpoints to form a new quadrilateral. What kind of quadrilateral is formed?

Geometry
Chapter 6 Resource Book

NAME _____ DATE _____

Reteaching with Practice

For use with pages 364–370

GOAL **Identify special quadrilaterals based on limited information and prove that a quadrilateral is a special type of quadrilateral, such as a rhombus or a trapezoid**

> **Ways to Prove a Shape is a Rhombus**
>
> 1. You can use the definition and show that the quadrilateral is a *parallelogram* that has four congruent sides. It is easier, however, to use the Rhombus Corollary and simply show that all four sides of the quadrilateral are congruent.
>
> 2. Show that the quadrilateral is a parallelogram and that the diagonals are perpendicular. *(Theorem 6.11)*
>
> 3. Show that the quadrilateral is a parallelogram and that each diagonal bisects a pair of opposite angles. *(Theorem 6.12)*

EXAMPLE *Proving a Quadrilateral is a Rhombus*

Show that *ABCD* is a rhombus.

SOLUTION

There are several ways of solving this problem.

Method 1 Use the Rhombus Corollary and show that all four sides of the quadrilateral are congruent.

$$AB = \sqrt{(1 - (-3))^2 + (3 - 1)^2} = \sqrt{20}$$

$$BC = \sqrt{(5 - 1)^2 + (1 - 3)^2} = \sqrt{20}$$

$$CD = \sqrt{(1 - 5)^2 + (-1 - 1)^2} = \sqrt{20}$$

$$DA = \sqrt{(-3 - 1)^2 + (1 - (-1))^2} = \sqrt{20}$$

So, because $AB = BC = CD = DA$, *ABCD* is a rhombus.

Method 2 Show that the quadrilateral is a parallelogram and that the diagonals are perpendicular.

Slope of $\overline{AB} = \dfrac{3 - 1}{1 - (-3)} = \dfrac{2}{4} = \dfrac{1}{2}$ Slope of $\overline{CD} = \dfrac{-1 - 1}{1 - 5} = \dfrac{-2}{-4} = \dfrac{1}{2}$

Slope of $\overline{BC} = \dfrac{1 - 3}{5 - 1} = \dfrac{-2}{4} = -\dfrac{1}{2}$ Slope of $\overline{DA} = \dfrac{1 - (-1)}{-3 - 1} = \dfrac{2}{-4} = -\dfrac{1}{2}$

So, quadrilateral *ABCD* is a parallelogram because opposite sides are parallel.

Slope of $\overline{AC} = \dfrac{1 - 1}{5 - (-3)} = \dfrac{0}{8} = 0$ (horizontal line)

Slope of $\overline{BD} = \dfrac{3 - (-1)}{1 - 1} = \dfrac{4}{0} =$ undefined   (vertical line)

So, because the diagonals are perpendicular (in this case, one is horizontal and the other is vertical—more generally, the slope of one would be

NAME _____ DATE _____

Reteaching with Practice

For use with pages 364–370

the negative reciprocal of the other's slope) and *ABCD* is a parallelogram, *ABCD* is a rhombus.

Exercises for Example 1

In Exercises 1 and 2, show that the quadrilateral with given vertices is a rhombus by using the methods demonstrated in Example 1.

1. $A(-4, -1)$, $B(-4, 2)$, $C(-1, 2)$, and $D(-1, -1)$. Use Method 1 from Example 1.

2. $E(-1, -2)$, $F(3, -1)$, $G(7, -2)$, and $H(3, -3)$. Use Method 2 from Example 1.

EXAMPLE 2 *Identifying Special Quadrilaterals*

For $P(0, 4)$, $Q(-4, 5)$, $R(-5, -1)$, and $S(-1, -2)$, what kind of quadrilateral is *PQRS*?

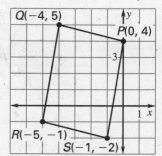

SOLUTION

Plot P, Q, R, and S in a coordinate plane.

Draw segments $\overline{PQ}$, $\overline{QR}$, $\overline{RS}$, and $\overline{SP}$.

PQRS looks like a rectangle. Begin by seeing if *PQRS* is a parallelogram.

Slope of $\overline{PQ} = \dfrac{5 - 4}{-4 - 0} = \dfrac{1}{-4} = -\dfrac{1}{4}$ Slope of $\overline{RS} = \dfrac{-2 - (-1)}{-1 - (-5)} = \dfrac{-1}{4} = -\dfrac{1}{4}$

Slope of $\overline{QR} = \dfrac{-1 - 5}{-5 - (-4)} = \dfrac{-6}{-1} = 6$ Slope of $\overline{SP} = \dfrac{4 - (-2)}{0 - (-1)} = \dfrac{6}{1} = 6$

So, because the slopes are equal, $\overline{PQ} \parallel \overline{RS}$ and $\overline{QR} \parallel \overline{SP}$. Therefore *PQRS* is a parallelogram. Next, see if the diagonals are congruent.

$$QS = \sqrt{(-1 - (-4))^2 + (-2 - 5)^2} = \sqrt{58}$$
$$PR = \sqrt{(-5 - 0)^2 + (-1 - 4)^2} = \sqrt{50}$$

The diagonals are not congruent, so *PQRS* is not a rectangle.

Thus, our conclusion about *PQRS* is that it is a parallelogram.

Exercises for Example 2

Given coordinates for P, Q, R, and S, what kind of quadrilateral is PQRS?

3. $P(-2, 1)$, $Q(-2, 3)$, $R(3, 6)$, $S(0, 1)$

4. $P(0, 0)$, $Q(4, 0)$, $R(3, 7)$, $S(1, 7)$

5. $P(-1, -3)$, $Q(4, -3)$, $R(4, 3)$, $S(-1, 3)$

NAME _____ DATE _____

Quick Catch-Up for Absent Students

For use with pages 364–370

The items checked below were covered in class on (date missed) _____

Lesson 6.6: Special Quadrilaterals

_____ **Goal 1:** Identify special quadrilaterals based on limited information. (p. 364)

Material Covered:

_____ Example 1: Identifying Quadrilaterals

_____ Example 2: Connecting Midpoints of Sides

_____ **Goal 2:** Prove that a quadrilateral is a special type of quadrilateral. (pp. 365–366)

Material Covered:

_____ Student Help: Look Back

_____ Example 3: Proving a Quadrilateral is a Rhombus

_____ Example 4: Identifying a Quadrilateral

_____ Example 5: Identifying a Quadrilateral

_____ Other (specify) _____

_____

Homework and Additional Learning Support

_____ Textbook (specify) pp. 367–370 _____

_____

_____ *Reteaching with Practice* worksheet (specify exercises)_____

_____ *Personal Student Tutor* for Lesson 6.6

NAME _____ DATE _____

Cooperative Learning Activity

For use with pages 364–370

GOAL **To investigate the connections between various points on a quadrilateral and the formation of special quadrilaterals**

Materials: 8.5 inch × 11 inch blank paper, pencil, ruler

Exploring Special Quadrilaterals

There are a number of relationships between special points in a quadrilateral that form special quadrilaterals. For example, joining the midpoints of the sides of any quadrilateral will form a parallelogram. There are two medians for each corner angle in a rectangle, one to the midpoint of each non-adjacent side. Connecting angles in a rectangle to the midpoints of other sides creates different shapes.

Instructions

1 On a blank piece of paper, mark the midpoint of each side of the paper.

2 From one corner of the paper, draw a line from the corner to the midpoint of the two non-adjacent sides (see figure). Repeat this procedure for each of the four corners of the paper, allowing each member of the group to complete one corner.

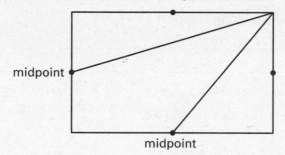

3 Repeat Step 2 for the octagon that is formed in the center of the paper, connecting each angle to the midpoints of the non-adjacent sides. Allow each member of your group to take a turn.

Analyzing the Results

1. Identify the special quadrilaterals that have been created during the activity.

2. How would the shape in the center of the paper change if the piece of paper was a square?

NAME _____ DATE _____

Real-Life Application: When Will I Ever Use This?

For use with pages 364–370

United States of America

After the Revolutionary War, the United States government had to decide how to organize the settlement of the land they acquired from England. Known as the Northwest Territory, this land was north of the Ohio River and east of the Mississippi River. In 1787 Congress passed the Northwest Ordinance, which established the standard by which much westward expansion took place. The ordinance described a plan by which territories could become states. When the population of a territory reached 60,000 and a state constitution was written and approved by Congress, statehood could be obtained.

The original Northwest Territory became the states of Ohio, Indiana, Illinois, Michigan, and Wisconsin, but a total of thirty-one states originated under the provisions of the Northwest Ordinance over several decades. Expansion was virtually continuous until 1912 when the last contiguous state, Arizona, was added. The last two of the current fifty states, Alaska and Hawaii, were not added until 1959.

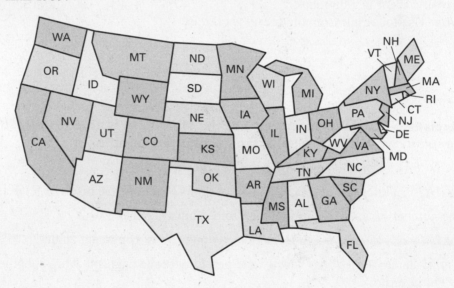

In Exercises 1–6, use the figure above. Which states (if any) appear to have the given shape?

1. Parallelogram   2. Rhombus   3. Rectangle

4. Square   5. Trapezoid   6. Kite

Challenge: Skills and Applications

For use with pages 364–370

1. Let *ABCD* be a quadrilateral with $\overline{AB} \cong \overline{BC}$, $\overline{CD} \cong \overline{DA}$, and $\overline{AB} \parallel \overline{CD}$. What type of quadrilateral is *ABCD*? Be as specific as possible, and write a two-column proof. (*Hint:* Begin your proof by drawing $\overline{AC}$.)

2. Let *EFGH* be a quadrilateral in which $\angle HEF$ and $\angle FGH$ are right angles and $\overline{EG}$ bisects both $\angle HEF$ and $\angle FGH$. What type of quadrilateral is *EFGH*? Be as specific as possible, and write a paragraph proof.

3. Write the key steps of a proof.
   Given: Quadrilateral *IJKL* with $\overline{IJ} \cong \overline{KL}$ and
   $\quad\quad\quad\angle IJK \cong \angle JKL$.

   Prove: $\overline{JK} \parallel \overline{LI}$
   (*Hint:* Draw additional segments as shown. One approach involves showing that $\triangle JKM$ and $\triangle ILM$ are isosceles.)

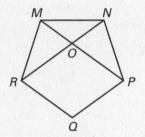

4. Refer to the diagram. Write a paragraph proof.
   Given: *MNPQR* is a regular pentagon.

   Prove: *OPQR* is a rhombus.
   (*Hint:* Use the result from the previous exercise.)

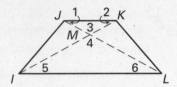

In Exercises 5–10, determine whether the statement is *true* or *false*. If it is true, explain why. If it is false, sketch a counterexample.

5. If *CDEF* is a kite, then *CDEF* is a convex polygon.

6. If *GHIJ* is a kite, then *GHIJ* is not a trapezoid. (*Hint:* Use the result of Exercise 1.)

7. The number of acute angles in a trapezoid is always either 1 or 2.

8. If *KLMN* is a kite, then *KLMN* has exactly one pair of congruent angles.

9. If quadrilateral *OPQR* has exactly one pair of opposite angles that are congruent, then *OPQR* is a kite.

10. In quadrilateral *STUV*, if $\overline{ST} \cong \overline{TU}$ and *T* is on the bisector of $\angle V$, then *STUV* is a kite.

TEACHER'S NAME _____ CLASS _____ ROOM _____ DATE _____

Lesson Plan

2-day lesson (See *Pacing the Chapter*, TE pages 318C–318D)   For use with pages 371–380

GOALS   1. **Find the areas of squares, rectangles, parallelograms, and triangles.**
            2. **Find the areas of trapezoids, kites, and rhombuses.**

State/Local Objectives _____

_____

✓ Check the items you wish to use for this lesson.

STARTING OPTIONS
_____ Homework Check: TE page 367: Answer Transparencies
_____ Warm-Up or Daily Homework Quiz: TE pages 372 and 370, CRB page 95, or Transparencies

TEACHING OPTIONS
_____ Motivating the Lesson: TE page 373
_____ Concept Activity: SE page 371; CRB pages 96–97 (Activity Support Master)
_____ Lesson Opener (Visual Approach): CRB page 98 or Transparencies
_____ Technology Activity with Keystrokes: CRB pages 99–101
_____ Examples: Day 1: 1–5: SE pages 373–375; Day 2: 6, SE page 375
_____ Extra Examples: Day 1: TE pages 373–375 or Transp.; Day 2: TE page 375 or Transp.
_____ Closure Question: TE page 375
_____ Guided Practice: SE page 376   Day 1: Exs. 1–13; Day 2: none

APPLY/HOMEWORK
Homework Assignment
_____ Basic   Day 1: 14–34; Day 2: 35–43, 50–52, 60, 61, 63–70; Quiz 3: 1–7
_____ Average   Day 1: 14–34; Day 2: 35–57, 60, 61, 63–70; Quiz 3: 1–7
_____ Advanced   Day 1: 14–34; Day 2: 35–70; Quiz 3: 1–7

Reteaching the Lesson
_____ Practice Masters: CRB pages 102–104 (Level A, Level B, Level C)
_____ Reteaching with Practice: CRB pages 105–106 or Practice Workbook with Examples
_____ Personal Student Tutor

Extending the Lesson
_____ Applications (Interdisciplinary): CRB page 108
_____ Challenge: SE page 379; CRB page 109 or Internet

ASSESSMENT OPTIONS
_____ Checkpoint Exercises: Day 1: TE pages 373–375 or Transp.; Day 2: TE page 375 or Transp.
_____ Daily Homework Quiz (6.7): TE page 380, or Transparencies
_____ Standardized Test Practice: SE page 379; TE page 380; STP Workbook; Transparencies
_____ Quiz (6.6–6.7): SE page 380

Notes _____

_____

_____

TEACHER'S NAME _____ CLASS _____ ROOM _____ DATE _____

Lesson Plan for Block Scheduling

1-day lesson (See *Pacing the Chapter*, TE pages 318C–318D)   For use with pages 371–380

GOALS   1. **Find the areas of squares, rectangles, parallelograms, and triangles.**
2. **Find the areas of trapezoids, kites, and rhombuses.**

State/Local Objectives _____

_____

✓ **Check the items you wish to use for this lesson.**

STARTING OPTIONS

____ Homework Check: TE page 367: Answer Transparencies
____ Warm-Up or Daily Homework Quiz: TE pages 372 and
        370, CRB page 95, or Transparencies

TEACHING OPTIONS

____ Motivating the Lesson: TE page 373
____ Concept Activity: SE page 371; CRB pages 96–97 (Activity Support Master)
____ Lesson Opener (Visual Approach): CRB page 98 or Transparencies
____ Technology Activity with Keystrokes: CRB pages 99–101
____ Examples 1–6: SE pages 373–375
____ Extra Examples: TE pages 373–375 or Transparencies
____ Closure Question: TE page 375
____ Guided Practice Exercises: SE page 376

APPLY/HOMEWORK

Homework Assignment

____ Block Schedule: 14–57, 63–70; Quiz 3: 1–7

Reteaching the Lesson

____ Practice Masters: CRB pages 102–104 (Level A, Level B, Level C)
____ Reteaching with Practice: CRB pages 105–106 or Practice Workbook with Examples
____ Personal Student Tutor

Extending the Lesson

____ Applications (Interdisciplinary): CRB page 108
____ Challenge: SE page 379; CRB page 109 or Internet

ASSESSMENT OPTIONS

____ Checkpoint Exercises: TE pages 373–375 or Transparencies
____ Daily Homework Quiz (6.7): TE page 380, or Transparencies
____ Standardized Test Practice: SE page 379; TE page 380; STP Workbook; Transparencies
____ Quiz (6.6–6.7): SE page 380

CHAPTER PACING GUIDE	
Day	**Lesson**
1	Assess Ch. 5; 6.1 (begin)
2	6.1 (end); 6.2 (begin)
3	6.2 (end); 6.3 (begin)
4	6.3 (end); 6.4 (begin)
5	6.4 (end); 6.5 (begin)
6	6.5 (end); 6.6 (all)
7	**6.7 (all)**
8	Review Ch. 6; Assess Ch. 6

Notes _____

_____

_____

NAME _____ DATE _____

WARM-UP EXERCISES

For use before Lesson 6.7, pages 371–380

1. $P = 2L + 2W$; Find P when $L = 3$ and $W = 5$.

2. $A = \pi r^2$; Find A for $\pi = 3.14$ and $r = 3$.

3. $A = \pi d$; Find d for $\pi = 3.14$ and $A = 21.98$.

4. $P = 4s$; Find P when $s = 10$.

5. $P = 2b + a$; Find b when $a = 5$ and $P = 23$.

DAILY HOMEWORK QUIZ

For use after Lesson 6.6, pages 364–370

1. For which special quadrilaterals is it *sometimes*, but *not always* true that the diagonals are congruent?

2. What kinds of quadrilaterals meet the conditions shown? *RSTU* is not drawn to scale.

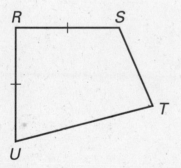

What kind of quadrilateral is *EFGH*? Justify your answer.

3. $E(-3, -2), F(-1, 1), G(2, 3), H(3, -3)$

4. $E(1, 4), F(4, 3), G(-1, -2), H(-2, 1)$

NAME _____ DATE _____

Activity Support Master

For use with page 371

Exploring Area of a Parallelogram

NAME _____ DATE _____

Activity Support Master

For use with page 371

Exploring Area of a Parallelogram

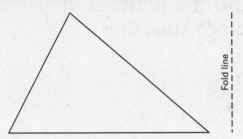

Fold line

Exploring Area of a Parallelogram

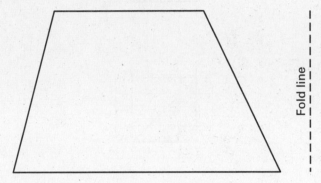

Fold line

NAME _____ DATE _____

Visual Approach Lesson Opener

For use with pages 372–380

Count the squares to find the area of the shaded figure. The dashed lines may help you to find the number of shaded squares without having to piece together partially shaded squares.

1.

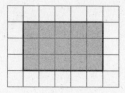

2.

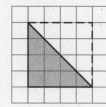

3.

4.

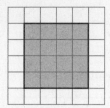

5.

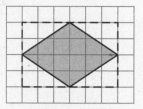

6.

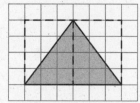

7.

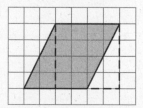

8.

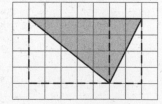

Geometry
Chapter 6 Resource Book

Lesson 6.7

NAME _____ DATE _____

Technology Activity

For use with pages 372–380

GOAL **To determine several properties about the diagonals of a kite and their relationship to the area**

Activity

❶ Construct kite *ABCD* (see figure). Use the software's grid feature and place vertex *A* at $(-1, 0)$, vertex *B* at $(0, 1)$, vertex *C* at $(2, 0)$, and vertex *D* at $(0, -1)$.

❷ Construct the diagonals of the kite, $\overline{AC}$ and $\overline{BD}$.

❸ Draw *E*, the point of intersection of $\overline{AC}$ and $\overline{BD}$.

❹ Measure the lengths of $\overline{AC}$, $\overline{BD}$, $\overline{AE}$, $\overline{EC}$, $\overline{BE}$, and $\overline{ED}$.

❺ Drag a vertex of kite *ABCD* and observe the results.

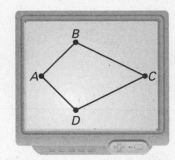

Exercises

1. Make a conjecture about the segment lengths of the diagonals.

2. Write an expression for the area of triangle *ABC*.

3. Write an expression for the area of triangle *ADC*.

4. Use the results of Exercises 2 and 3 to write an expression for the area of this kite. Compare your result to Theorem 6.24.

Technology Activity Keystrokes
For use with pages 372–380

TI-92

1. Construct kite *ABCD*.

[F8] 9 (Set Coordinate Axes to RECTANGULAR and Grid to ON.) [ENTER]

[F3] 4 (Move cursor to $(-1, 0)$ and prompt says, "POINT ON. . . .") [ENTER] 2 (Move cursor to $(0, 1)$ and prompt says, "POINT ON. . . .") [ENTER] 2 (Move cursor to $(2, 0)$ and prompt says, "POINT ON. . . .") [ENTER] 2 (Move cursor to $(0, -1)$ and prompt says, "POINT ON. . . .") [ENTER] 2 (Move cursor to vertex at $(-1, 0)$.) [ENTER]

Label the vertices.

[F7] 4 (Move cursor to vertex at $(-1, 0)$.) [ENTER] *A* [ENTER] (Move cursor to vertex at $(0, 1)$.) [ENTER] *B* [ENTER] (Move cursor to vertex at $(2, 0)$.) [ENTER] *C* [ENTER] (Move cursor to vertex at $(0, -1)$.) [ENTER] *D* [ENTER]

Turn off the axes and the grid.

[F8] 9 (Set Coordinate Axes to OFF and Grid to OFF.) [ENTER]

2. Construct the diagonals of the kite, $\overline{AC}$ and $\overline{BD}$.

[F2] 5 (Move cursor to *A*.) [ENTER] (Move cursor to *C*.) [ENTER] (Move cursor to *B*.) [ENTER] (Move cursor to *D*.) [ENTER]

3. Draw *E*, the point of intersection of $\overline{AC}$ and $\overline{BD}$.

[F2] 3 (Move cursor to intersection of $\overline{AC}$ and $\overline{BD}$.) [ENTER] *E*

4. Measure $\overline{AC}$, $\overline{BD}$, $\overline{AE}$, $\overline{EC}$, $\overline{BE}$, and $\overline{ED}$.

[F6] 1 (Move cursor to $\overline{AC}$.) [ENTER] (Move cursor to $\overline{BD}$.) [ENTER] (Move cursor to *A*.) [ENTER] (Move cursor to *E*.) [ENTER] [ENTER] (Move cursor to *C*.) [ENTER] (Move cursor to *B*.) [ENTER] (Move cursor to *E*.) [ENTER] [ENTER] (Move cursor to *D*.) [ENTER]

5. Use the drag key 🖐 and the cursor pad to drag a vertex of kite *ABCD*.

Geometry
Chapter 6 Resource Book

NAME _____ DATE _____

Technology Activity Keystrokes

For use with pages 372–380

SKETCHPAD

1. Turn on the axes and the grid by choosing **Snap To Grid** from the **Graph** menu.

   Choose the segment straightedge tool.

   Draw a segment from $(-1, 0)$ to $(0, 1)$.

   Draw a segment from $(0, 1)$ to $(2, 0)$.

   Draw a segment from $(2, 0)$ to $(0, -1)$.

   Draw a segment from $(0, -1)$ to $(-1, 0)$. Label the vertices.

   Choose the text tool. Label vertex $(-1, 0)$ A, label vertex $(0, 1)$ B, label vertex $(2, 0)$ C, and label vertex $(0, -1)$ D.

   Turn off the axes and the grid.

   Choose **Hide Axes** from the **Graph** menu.

   Choose **Hide Grid** from the **Graph** menu.

2. Choose the segment straightedge tool and construct the diagonals of the kite, $\overline{AC}$ and $\overline{BD}$.

3. Choose the point tool and draw E, the point of intersection of $\overline{AC}$ and $\overline{BD}$. Relabel the point using the text tool, if necessary.

4. Measure the lengths of $\overline{AC}$, $\overline{BD}$, $\overline{AE}$, $\overline{EC}$, $\overline{BE}$, and $\overline{ED}$.

   Choose the selection arrow tool and select $\overline{AC}$. Then hold down the shift key, select $\overline{BD}$, and choose **Length** from the **Measure** menu.

   To measure $\overline{AE}$, select A, hold down the shift key, select E, and choose **Distance** from the **Measure** menu.

   Repeat this process for $\overline{EC}$, $\overline{BE}$, and $\overline{ED}$, making sure that you have only selected the endpoints of the segment you are measuring.

5. Use the translate selection arrow tool to drag a vertex of kite $ABCD$.

Practice A

For use with pages 372–380

Find the area of the polygon.

1.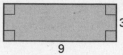

2.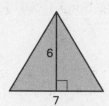

3.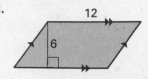

4.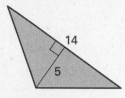

5.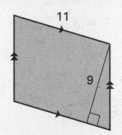

6.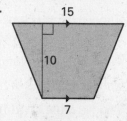

7.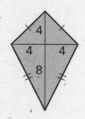

8.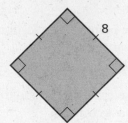

9.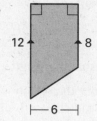

Use the Pythagorean Theorem to find the area of the polygon.

10.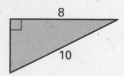

11.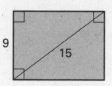

12.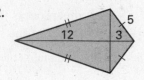

13. The garage roof shown is made from two isosceles trapezoids and two isosceles triangles. Find the area of the entire roof.

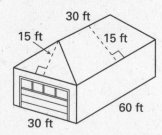

14. The Millers have two gardens as shown below. The shaded region represents the lawn that needs to be fertilized. Find the area of the lawn.

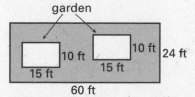

Find the area of the polygon.

1.

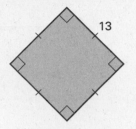

13

2.

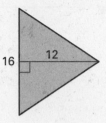

16   12

3.

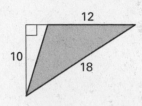

12
10
18

4.

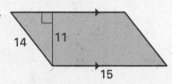

14   11
15

5.

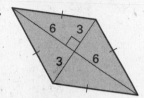

6   3
3   6

6.

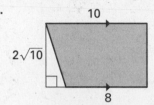

10
$2\sqrt{10}$
8

7.

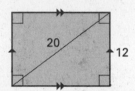

20   12

8.

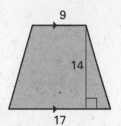

9
14
17

9.

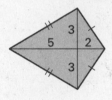

3
5   2
3

The quadrilateral has an area of 64 square units. Find the value of *x*.

10.

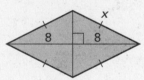

x
8   8

11.

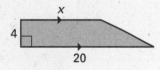

x
4
20

12.

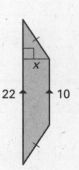

x
22   10

Find the areas of the shaded and unshaded regions.

13.

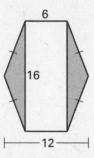

6
16
12

14.

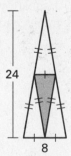

24
8

15.

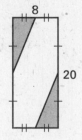

8
20

NAME _____ DATE _____

Practice C
For use with pages 372–380

Find the area of the polygon.

1.
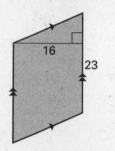
7

2.
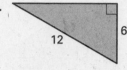
12
6

3.
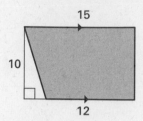
15
10
12

4.
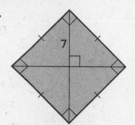
16
23

5.
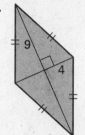
9
4

6.
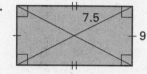
7.5
9

7.
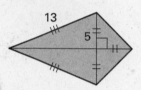
13
5

8.
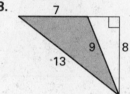
7
9
8
13

9.
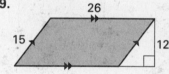
26
15
12

Find the area of the quadrilateral.

10.
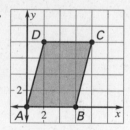

11.
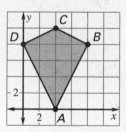

12.
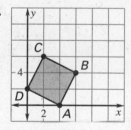

13. You want to fill in the space between the handrail and steps on your front porch for the winter months. How much plywood would you need?

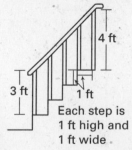
4 ft
3 ft
1 ft
Each step is
1 ft high and
1 ft wide

14. You are making a kite. The frame is to be made from two pieces of balsa wood, one measuring 64 inches and the other 38 inches. If you buy 1 square yard of material, will you have enough to piece together the covering for the kite? Explain.

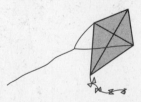

Lesson 6.7

NAME _____ DATE _____

Reteaching with Practice

For use with pages 372–380

GOAL Find the areas of squares, rectangles, parallelograms, and triangles and find the areas of trapezoids, kites, and rhombuses

---

Postulate 22 Area of a Square Postulate
The area of a square is the square of the length of its side.

Postulate 23 Area Congruence Postulate
If two polygons are congruent, then they have the same area.

Postulate 24 Area Addition Postulate
The area of a region is the sum of the areas of its nonoverlapping parts.

Theorem 6.20 Area of a Rectangle
The area of a rectangle is the product of its base and height.

Theorem 6.21 Area of a Parallelogram
The area of a parallelogram is the product of a base and its corresponding height.

Theorem 6.22 Area of a Triangle
The area of a triangle is one half the product of a base and its corresponding height.

Theorem 6.23 Area of a Trapezoid
The area of a trapezoid is one half the product of the height and the sum of the bases.

Theorem 6.24 Area of a Kite
The area of a kite is one half the product of the lengths of its diagonals.

Theorem 6.25 Area of a Rhombus
The area of a rhombus is equal to one half the product of the lengths of the diagonals.

EXAMPLE 1 *Using the Area Theorems*

a. Find the area of △ABC.

b. Find the area of ▱DEFG.

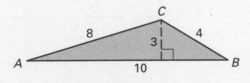

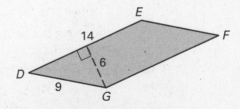

c. Find the area of trapezoid WXYZ.

d. Find the area of kite ABCD. $DB = 6$, $AC = 8$

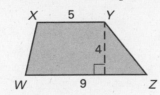

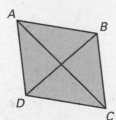

Lesson 6.7

Reteaching with Practice

For use with pages 372–380

SOLUTION

a. Any one of the three sides of the triangle can be used as the base, b. Since the height of 3, which corresponds to the base $\overline{AB}$, is given, use $\overline{AB}$ for b.

$$\text{Area} = \frac{1}{2}bh = \frac{1}{2}(10)(3) = 15 \text{ square units}$$

b. We use $\overline{DE}$ as the base because the corresponding height of 6 is given.
Area $= bh = (14)(6) = 84$ square units

c. Area $= \frac{1}{2}h(b_1 + b_2) = \frac{1}{2}(4)(5 + 9) = 28$ square units

d. Area $= \frac{1}{2}d_1d_2 = \frac{1}{2}(6)(8) = 24$ square units

Exercises for Example 1

Find the area of the polygon.

1.
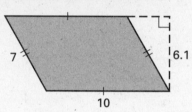

2.
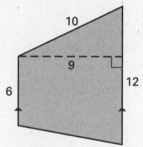

3.
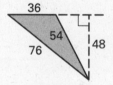

4.
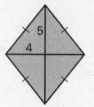

5.
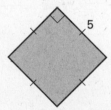

6.
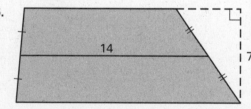

NAME _____ DATE _____

Quick Catch-Up for Absent Students

For use with pages 371–380

The items checked below were covered in class on (date missed) _____

Activity 6.7: Areas of Quadrilaterals (p. 371)

_____ **Goal:** Determine how the areas of a rectangle, parallelogram, triangle, and trapezoid are related to each other.

Lesson 6.7: Areas of Triangles and Quadrilaterals

_____ **Goal 1:** Find the areas of squares, rectangles, parallelograms, and triangles. (pp. 372–373)

Material Covered:

_____ Student Help: Study Tip

_____ Example 1: Using the Area Theorems

_____ Example 2: Finding the Height of a Triangle

_____ Student Help: Study Tip

_____ Example 3: Finding the Height of a Triangle

_____ **Goal 2:** Find the areas of trapezoids, kites, and rhombuses. (pp. 374–375)

Material Covered:

_____ Student Help: Look Back

_____ Example 4: Finding the Area of a Trapezoid

_____ Example 5: Finding the Area of a Rhombus

_____ Student Help: Study Tip

_____ Example 6: Finding Areas

_____ Other (specify) _____

_____

Homework and Additional Learning Support

_____ Textbook (specify) _pp. 376–380_____

_____

_____ *Reteaching with Practice* worksheet (specify exercises)_____

_____ *Personal Student Tutor* for Lesson 6.7

NAME _____ DATE _____

Interdisciplinary Application

For use with pages 372–380

Bluebird Houses

WOODSHOP   A species once readily seen throughout the United States, the
bluebird population has slowly declined over the past century. Destruction of the
bluebird's natural habitat has made nesting difficult. Because bluebirds depend
on abandoned woodpecker holes or rotting trees for shelter, a manmade
birdhouse can be a great alternative.

A bluebird house should have a 5 inch by 5 inch floor, a height ranging from
8 inches to 12 inches, and an opening of 1.5 inches. Bluebirds like open areas,
such as fields, large lawns, cemeteries, parks, and golf courses. Hanging a
bluebird house on a post or pole about 4 or 5 feet above the ground will
probably attract the birds and offer them protection from predators.

In Exercises 1–5, use the following information.

Your woodshop class is making bluebird houses for a final project. A model of
the birdhouse is shown below. The roof consists of four congruent triangles, the
sides are made of four congruent isosceles trapezoids, and the base includes four
rectangles and a square bottom.

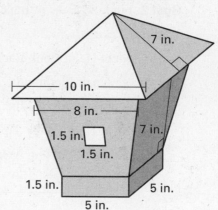

1. Find the area of material needed to
   make the roof.

2. Find the area of the entrance hole.

3. Find the area of material needed to
   make the sides of the birdhouse,
   excluding the entrance hole.

4. Find the area of the material needed
   to construct the base of the birdhouse.

5. What is the total amount of material
   needed for the entire birdhouse?

Challenge: Skills and Applications

For use with pages 372–380

1. In the figure shown, D, E, F, G, H, and I are midpoints of segments. If the area of $\triangle GHI$ is 12, what is the area of $\triangle ABC$?

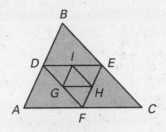

2. Use the diagram to write a plan for a proof of Theorem 6.23 (Area of a Trapezoid).

   Given: $JKLM$ is a trapezoid with bases $\overline{JK}$ and $\overline{ML}$.

   Prove: The area of $JKLM$ is $\frac{1}{2}h(b_1 + b_2)$.

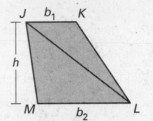

3. Write a paragraph proof showing that $PQRS$ is a parallelogram if and only if the diagonals divide $PQRS$ into four triangles of equal area.

   a. Given: $PQRS$ is a parallelogram.

   Prove: $\triangle OPQ$, $\triangle OQR$, $\triangle ORS$, and $\triangle OSP$ all have the same area.

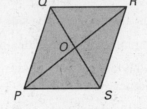

   b. Given: $\triangle OPQ$, $\triangle OQR$, $\triangle ORS$, and $\triangle OSP$ all have the same area.

   Prove: $PQRS$ is a parallelogram.

4. Find the area of a trapezoid whose height is 5 and whose midsegment has length 4.

In Exercises 5–7, use the following information to find the area of the triangle.

If the side lengths of a triangle are known, the area can be determined using *Heron's Formula*, even if the altitude is not known. Let a, b, and c represent the lengths of the three sides, and let s represent one-half the perimeter. Then the area of the triangle is $\sqrt{s(s - a)(s - b)(s - c)}$.

5.
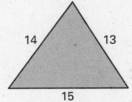

6.
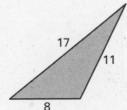

7.
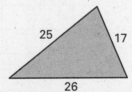

Lesson 6.7

NAME _____ DATE _____

Chapter Review Games and Activities

For use after Chapter 6

Complete the following number crossword puzzle. All answers are positive integers.

Down

1. In a parallelogram, the measure of an angle that is consecutive to an angle of measure 6°.

3. The length of the midsegment of a trapezoid with bases length of 3664 and 2500.

5. The sum of the measures of the interior angles of a quadrilateral.

7. The length of the side of a square with area 27,709,696.

9. The area of a kite with diagonal lengths of 8 and 147.

10. Twenty times the area of a rectangle that has a diagonal length of 5 and a height of 3.

Across

2. In a rhombus, the length of the side opposite a side with length 736.

4. 100 times the number of sides of a parallelogram.

6. The sum of the lengths of a kite with the length of one side 211 and the length of another side 315.

8. The length of the remaining base of a trapezoid with height 16, base 4896, and area 75,264.

11. The height of a triangle with a base length of 4 and an area of 1620.

Chapter Test B

For use after Chapter 6

Decide whether the figure is a polygon.

1.   2.   3.   4.

State whether the figure is _convex_ or _concave_.

5.   6.   7.   8.

Decide whether the statement is _always_, _sometimes_, or _never_ true.

9. A rhombus is a square.

10. A rectangle is a parallelogram.

11. A trapezoid is a parallelogram.

12. A parallelogram is a rectangle.

Find the values of _x_.

13.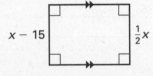

14.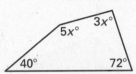

15.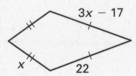

16.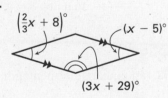

1.	_____
2.	_____
3.	_____
4.	_____
5.	_____
6.	_____
7.	_____
8.	_____
9.	_____
10.	_____
11.	_____
12.	_____
13.	_____
14.	_____
15.	_____
16.	_____
17.	_____
18.	_____
19.	_____
20.	_____

Decide if you are given enough information to prove that the quadrilateral is a parallelogram.

17. One pair of opposite sides are congruent.

18. Two pairs of opposite angles are congruent.

19. All pairs of consecutive angles are congruent.

20. Diagonals are perpendicular.

Chapter Test B

For use after Chapter 6

What special type of quadrilateral is shown?

21.

22.

23.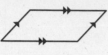

Tell whether the polygon is best described as *equiangular*, *equilateral*, *regular*, or *none of these*.

24.

25.

26.

27.

Find the area of the quadrilateral.

28.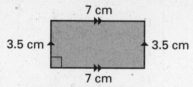

29.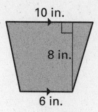

30.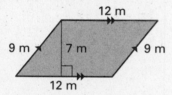

31.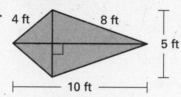

Draw a figure that fits the description.

32. an equilateral quadrilateral

33. an equiangular pentagon

34. a regular quadrilateral

35. a concave hexagon

21.	_____
22.	_____
23.	_____
24.	_____
25.	_____
26.	_____
27.	_____
28.	_____
29.	_____
30.	_____
31.	_____
32.	See left.
33.	See left.
34.	See left.
35.	See left.

NAME _____ DATE _____

Chapter Test C

For use after Chapter 6

Decide whether the figure is a polygon.

1.   2. 3. 4.

State whether the figure is *convex* or *concave*.

5. 6. 7. 8.

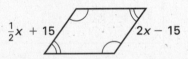

Decide whether the statement is *always, sometimes,* or *never* true.

9. A rhombus is a quadrilateral.

10. A rectangle is a trapezoid.

11. A trapezoid is an isosceles trapezoid.

12. A parallelogram is a rectangle.

Find the values of *x*.

13. $\frac{1}{2}x + 15$ $2x - 15$

14. 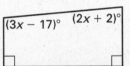 $(3x - 17)°$ $(2x + 2)°$

15. 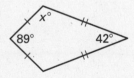 $x°$ $89°$ $42°$

16. 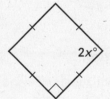 $2x°$

Decide if you are given enough information to prove that the quadrilateral is a parallelogram.

17. One angle is a right angle

18. Two pairs of opposite sides are congruent.

19. One pair of consecutive angles are congruent.

20. Diagonals are perpendicular and congruent.

21. Opposite sides are parallel and congruent.

22. All four sides are congruent.

| 1. _____ |
| 2. _____ |
| 3. _____ |
| 4. _____ |
| 5. _____ |
| 6. _____ |
| 7. _____ |
| 8. _____ |
| 9. _____ |
| 10. _____ |
| 11. _____ |
| 12. _____ |
| 13. _____ |
| 14. _____ |
| 15. _____ |
| 16. _____ |
| 17. _____ |
| 18. _____ |
| 19. _____ |
| 20. _____ |
| 21. _____ |
| 22. _____ |

Review and Assess

Chapter Test C

For use after Chapter 6

What special type of quadrilateral is shown?

23.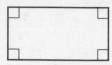

24.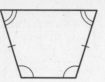

25.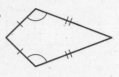

Tell whether the polygon is best described as *equiangular,*
equilateral, regular, **or** *none of these*.

26.

27.

28.

29.

Find the area of the quadrilateral.

30.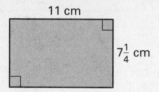

11 cm

$7\frac{1}{4}$ cm

31.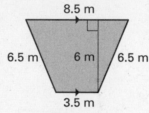

8.5 m

6.5 m   6 m   6.5 m

3.5 m

32.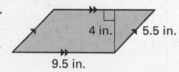

4 in.   5.5 in.

9.5 in.

33.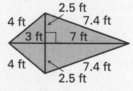

2.5 ft   7.4 ft
4 ft
3 ft   7 ft
4 ft   7.4 ft
2.5 ft

23.	_____
24.	_____
25.	_____
26.	_____
27.	_____
28.	_____
29.	_____
30.	_____
31.	_____
32.	_____
33.	_____
34.	See left.
35.	See left.
36.	See left.
37.	See left.

Draw a figure that fits the given description.

34. an equiangular quadrilateral

35. an equilateral pentagon
that is not equiangular

36. a regular decagon

37. a nonconvex quadrilateral

Review and Assess

NAME _____ DATE _____

SAT/ACT Chapter Test

For use after Chapter 6

1. In quadrilateral $ABCD$, $\overline{AB} \parallel \overline{CD}$ and $\angle A$ and $\angle B$ are supplementary. Which statements are true?

   I.   Quadrilateral $ABCD$ is regular.

   II. Quadrilateral $ABCD$ is a rectangle.

   III. Quadrilateral $ABCD$ is a rhombus.

   Ⓐ I only   Ⓑ I, II only

   Ⓒ II, III only   Ⓓ I, II, and III

   Ⓔ none of these

2. Find the area of a triangle with vertices $A(0, 2)$, $B(8, 2)$, and $C(4, -3)$.

   Ⓐ 17   Ⓑ 20   Ⓒ 15

   Ⓓ 19   Ⓔ 18

3. Find the value of x.

   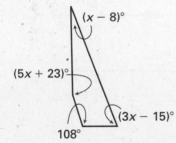

   Ⓐ 21
   Ⓑ 25
   Ⓒ 28
   Ⓓ 31
   Ⓔ 34

4. What are the values of the variables in quadrilateral $MNOP$?

   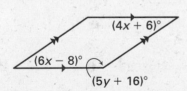

   Ⓐ $x = 4, y = 19$ Ⓑ $x = 6, y = 19$
   Ⓒ $x = 5, y = 27$ Ⓓ $x = 3, y = 32$
   Ⓔ $x = 7, y = 26$

5. $NPQR$ is a trapezoid and $ST = 24$. Find the value of x.

   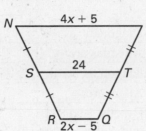

   Ⓐ 6
   Ⓑ 10
   Ⓒ 8
   Ⓓ 9
   Ⓔ 11

6. Find the area of a parallelogram with vertices $A(-4, 2)$, $B(1, 6)$, $C(15, 6)$, and $D(10, 2)$.

   Ⓐ 48 square units   Ⓑ 56 square units
   Ⓒ 60 square units   Ⓓ 51 square units
   Ⓔ 70 square units

7. Choose the statement that is true about a kite.

   Ⓐ Only one pair of opposite angles are congruent.

   Ⓑ Opposite sides are congruent.

   Ⓒ Diagonals bisect each other.

   Ⓓ Diagonals are congruent.

   Ⓔ None of these are true.

8. What special type of quadrilateral has the vertices $F(-6, -2)$, $G(1, -2)$, $H(-6, -5)$, and $I(1, -5)$?

   Ⓐ rectangle   Ⓑ square
   Ⓒ parallelogram Ⓓ rhombus
   Ⓔ kite

9. $DEFG$ is a trapezoid and $HI = 15.5$. Find the value of x.

   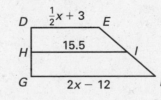

   Ⓐ 13
   Ⓑ 12
   Ⓒ 15
   Ⓓ 14
   Ⓔ 16

Review and Assess

JOURNAL 1. Copy the chart. Write *yes* in the box if the shape *always* has the given property.

Property		Rectangle	Rhombus	Square	Kite	Trapezoid
Diagonals are ≅	?	?	?	?	?	?
All sides are ≅	?	?	?	?	?	?
Diagonals are ⊥	?	?	?	?	?	?
All ∠s are ≅	?	?	?	?	?	?
Exactly 1 pair of opp. sides are ∥	?	?	?	?	?	?
Exactly 1 pair of opp. ∠s are ≅	?	?	?	?	?	?

MULTI-STEP PROBLEM 2. In the diagram at the right, *ADEH* is an isosceles trapezoid, *BCEH* is a rectangle, *EFGH* is a parallelogram, $GH = 8$, $GF = 24$, $FE = \frac{1}{2}ED$, $BC = \frac{1}{2}AD$, and $m\angle BAH = 43°$.

a. $m\angle HBC =$ __?__

b. $m\angle FDC =$ __?__

c. $m\angle HEF =$ __?__

d. $m\angle EFG =$ __?__

e. $BC =$ __?__

f. $AD =$ __?__

g. $ED =$ __?__

h. $FD =$ __?__

i. $AH =$ __?__

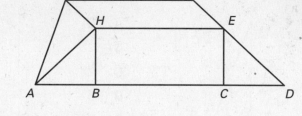

3. *Critical Thinking* Use the diagram from Exercise 2.

   a. Add the midpoint of $\overline{AH}$ and label the midpoint *X*. Add the midpoint of $\overline{DE}$ and label the midpoint *Y*. Find the length of $\overline{XY}$.

   b. Find the area of trapezoid *AGFD*. The height of trapezoid *AGFD* is 16.5 units.

4. *Writing* There are six ways to prove a shape is a parallelogram. Describe five of them.

Alternative Assessment Rubric

For use after Chapter 6

JOURNAL
SOLUTION

1. Complete answers should include:

Property		Rectangle	Rhombus	Square	Kite	Trapezoid
Diagonals are ≅		yes		yes		
All sides are ≅			yes	yes		
Diagonals are ⊥			yes	yes	yes	
All ⧌ are ≅		yes		yes		
Exactly 1 pair of opp. sides are ‖						yes
Exactly 1 pair of opp. ⧌ are ≅					yes	

MULTI-STEP
PROBLEM
SOLUTION

2. a. 90°   **b.** 43°   **c.** 43°

d. 137°   **e.** 24 units   **f.** 48 units

g. 16 units   **h.** 24 units   **i.** 16 units

3. a. 36 units

b. 594 square units

4. Answers may vary. Answers should include five of the six ways to prove a shape is a parallelogram.

MULTI-STEP
PROBLEM
RUBRIC

4   Students answer all parts of the problem correctly, showing their work. Students write five correct methods of proof. Students' explanations are clear and easy to follow.

3   Students answer the problem correctly. Students write five methods of proof. Students may have one incorrect method. Students' explanations are clear.

2   Students answer the problem, but may have some mathematical errors. Students may write less than five methods of proof. Students may have more than one incorrect method. Students' explanations do not match the methods to prove a shape is a parallelogram.

1   Students' answers are incomplete. Students do not write any methods of proof. Most of the methods are incorrect. Students' explanations are not clear and do not match the methods to prove a shape is a parallelogram.

Review and Assess

NAME _____ DATE _____

Project: "Puzzling" Shapes

For use with Chapter 6

OBJECTIVE Use a tangram puzzle set to create polygons and analyze their measurements.

MATERIALS ruler, thin cardboard or construction paper, scissors, and a small plastic bag; or a set of tangrams

INVESTIGATION Tangrams are considered to be an ancient Chinese puzzle. A tangram puzzle set can be made by cutting a square into the pieces indicated in the diagram.

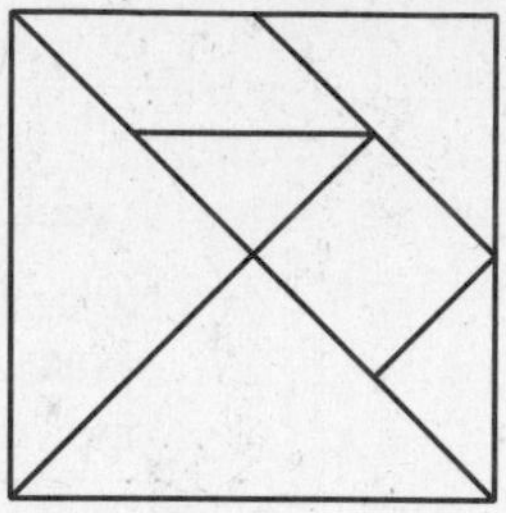

Exploring Tangrams

1. What do you notice about each of the following?

   a. the two large right triangles

   b. the legs of the medium right triangle

   c. the longer sides of the parallelogram

   d. the sides of the small square and the legs of the small right triangles

2. Create your own tangram puzzle set. Draw a 4-inch by 4-inch square and the remaining line segments as shown in the diagram above.

3. What are the polygons with the fewest sides and with the most sides that you can build using all seven pieces? Name your polygons based on the number of sides they have. Draw a sketch to support your answers.

4. Is it possible to form each of the shapes listed below using exactly two tangram pieces? three pieces? four pieces? five pieces? six pieces? seven pieces? Draw a small sketch of each one you formed, showing the pieces you used.

   a. square

   b. rectangle that is not a square

   c. parallelogram that is not a rectangle

   d. trapezoid

Analyzing Measurements

5. Find the area of each tangram puzzle piece when the area of the

   a. small triangle is 1 square unit?

   b. small triangle is $\frac{1}{2}$ square unit?

   c. small triangle is $\frac{1}{4}$ square unit?

   d. small square is 1 square unit?

   e. small square is $\frac{1}{2}$ square unit?

   f. small square is $\frac{1}{4}$ square unit?

   g. entire square is 1 square unit?

   h. entire square is 16 square units?

PRESENT YOUR RESULTS Your project report should include all of your sketches and measurement data. Also, include a written summary of your work showing calculations. Discuss how you approached the problems and how successful you were in finding an approach that helped you find as many polygons as possible. What techniques did you use to organize your work?

ANSWERS

Chapter Support

Parent Guide

6.1: pentagon; yes **6.2:** $\overline{AB} \cong \overline{DC}$, $\overline{AD} \cong \overline{BC}$, $\angle ADC$ and $\angle ABC$ **6.3:** *Sample answer:* make sure $AB = DC$ and $AD = BC$; Theorem 6.6 (See p. 338.) **6.4:** make sure $AC = BD$; Theorem 6.13 (See p. 349.) **6.5:** $WX = XY = 5$ m, $WZ = YZ \approx 6.71$ m **6.6:** always: rectangle, square, isosceles trapezoid; sometimes: parallelogram, rhombus **6.7:** 14 square units

Prerequisite Skills Review

1. Corresponding Angles Postulate
2. Alternate Exterior Angles Theorem
3. Alternate Interior Angles Theorem
4. Linear Pair Postulate
5. Consecutive Interior Angles Theorem
6. Vertical Angles Theorem **7.** SAS Postulate
8. SSS Postulate **9.** ASA Postulate
10. HL Theorem **11.** none
12. AAS Theorem
13. 11.66 units, $m = \frac{3}{5}$, $M = (4, 1)$
14. 4.47 units, $m = -\frac{1}{2}$, $M = (4, 6)$
15. 6.40 units, $m = \frac{4}{5}$, $M = \left(-\frac{5}{2}, 2\right)$
16. 8.49 units, $m = 1$, $M = (5, -7)$
17. 5 units, $m = -\frac{3}{4}$, $M = \left(2, \frac{3}{2}\right)$
18. 14.14 units, $m = \frac{1}{7}$, $M = (-1, 5)$

Strategies for Reading Mathematics

1. $\angle 1 \cong \angle 2$; it is written in the given statement.
2. $\overline{AB} \parallel \overline{DC}$; by the Alternate Interior Angles Converse Theorem.
3. using the Consecutive Interior Angles Theorem

4.

Statements	Reasons
1. $\angle 1 \cong \angle 2$	1. Given
2. $\overline{AB} \parallel \overline{DC}$	2. Alternate Interior Angles Converse Theorem
3. $\angle ABC$ and $\angle BCD$ are supplementary.	3. Consecutive Interior Angles Theorem

Lesson 6.1

Warm-Up Exercises

1. $180°$ **2.** 31 **3.** equilateral **4.** equiangular

Daily Homework Quiz

1. > **2.** > **3.** = **4.** >

Lesson Opener

Allow 5 minutes.

1. One of four children born at one birth.
2. Happening once every eight years. **3.** An athletic contest that consists of five events for each participant. **4.** Having three legs or feet. **5.** A person between ninety and one hundred years of age. **6.** A six-pointed star.
7. A solid geometric figure with ten plane faces.
8. A polygon with seven sides and seven angles.
9. A line of verse consisting of five metrical feet.

Practice A

1. yes **2.** no; curved edges **3.** yes
4. hexagon; convex **5.** dodecagon; concave
6. heptagon; concave **7.** hexagon
8. *NOPQRM*; *OPQRMN* **9.** $\overline{MQ}$, $\overline{MP}$, $\overline{MO}$
10. $\angle M$, $\angle O$ **11.** equilateral **12.** equiangular
13. regular **14.** 107 **15.** 30 **16.** 12

Practice B

1. no; curved edges **2.** yes
3. no; curved edges **4.** hexagon; concave
5. heptagon; convex **6.** heptagon; concave

Lesson 6.1 *continued*

7. heptagon   **8.** *BCDEFGA*; *CDEFGAB*

9. $\overline{EG}, \overline{EA}, \overline{EB}, \overline{EC}$ **10.** $\angle F, \angle E, \angle D, \angle C$

11. 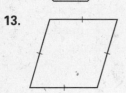   **12.**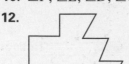

13. 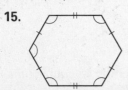   **14.**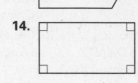

15. 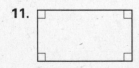   **16.** 55   **17.** 21   **18.** 17

Practice C

1. no; a vertex is shared by more than 2 sides

2. yes   **3.** no; curved edges

4. quadrilateral; convex   **5.** decagon; concave

6. heptagon; concave

7.   **8.**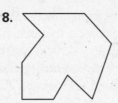

9.   **10.**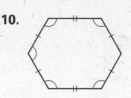

11.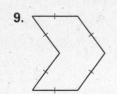

12. 43   **13.** 18   **14.** 13   **15.** true   **16.** false
17. true   **18.** false   **19.** false   **20.** false

Reteaching with Practice

1. yes   **2.** No, the one side of the figure is not a segment.   **3.** yes   **4.** No, two of the sides only intersect one other side.   **5.** convex

6. concave   **7.** concave   **8.** 90   **9.** 36

Interdisciplinary Application

1.

Polygon	Number of sides	Sum of the interior angles
Triangle	3	180°
Quadrilateral	4	360°
Pentagon	5	540°
Hexagon	6	720°
Heptagon	7	900°
Octagon	8	1080°

2. The total measure of the interior angles increases by 180°.   **3.** $S = (n - 2) \cdot 180°$, where n is the number of sides of the polygon.

4. Pentagon: $S = (5 - 2) \cdot 180° = 540°$

Hexagon: $S = (6 - 2) \cdot 180° = 720°$

Heptagon: $S = (7 - 2) \cdot 180° = 900°$

Octagon: $S = (8 - 2) \cdot 180° = 1080°$

5. $S = (100 - 2) \cdot 180° = 17,640°$

Challenge: Skills and Applications

1. 7   **2.** $-10, 11$ **3.** -12

4. $-12 < x < 20$ **5.** $x = 15, y = -20$

6. $x = 24, y = 6$ **7.** 540°   **8.** 720°   **9.** 1080°

10. true   **11.** false; *Sample answer:*

Lesson 6.2

Warm-Up Exercises

1. Transitive Prop. of Cong.

2. Symmetric Prop. of Cong.

3. Substitution prop. of equality

Daily Homework Quiz

1. nonagon; concave   **2.** Students' drawings should show a rhombus.   **3.** 113°   **4.** 15

Lesson 6.2 *continued*

Lesson Opener

Allow 10 minutes.

1. *Sample answer*:

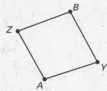

2. Use toothpicks of different lengths; change the angle at which the midpoints are aligned; descriptions will vary.

3. *Sample answer*:

By the lengths of the toothpicks; they are diagonals of *AYBZ*; they bisect each other.   **4.** *Sample conjectures:* Opposite sides are parallel; opposite sides are congruent; opposite angles are congruent; consecutive angles are supplementary; diagonals bisect each other. Figure *AYBZ* is a parallelogram.

Practice A

1. yes   **2.** No; only one pair of opposite sides are parallel.   **3.** No; it is a hexagon.

4. $\overline{PO}$; opposite sides of $\square$ are $\cong$.

5. $\overline{OP}$; opposite sides of $\square$ are parallel.

6. $\overline{PM}$; opposite sides of $\square$ are $\cong$.

7. $\angle ONM$; opposite $\angle$s of $\square$ are $\cong$.

8. $\overline{QN}$; diagonals of $\square$ bisect each other.

9. $\overline{QO}$; diagonals of $\square$ bisect each other.

10. $\angle PQO$; vertical $\angle$s are $\cong$.

11. $\angle PNM$; Alternate Interior $\angle$s Converse.

12. 16; opposite sides of $\square$ are $\cong$.

13. 10; opposite sides of $\square$ are $\cong$.

14. 8; diagonals of $\square$ bisect each other.

15. 16; diagonals of $\square$ bisect each other.

16. 28°; alternate interior $\angle$s are $\cong$.

17. 96°; consecutive $\angle$s of $\square$ are supplementary.

18. 84°; opposite $\angle$s of $\square$ are $\cong$.

19. 68°; opposite $\angle$s of $\square$ are $\cong$ and subtraction.

20. $x = 16, y = 6$ **21.** $x = 2, y = 9$

22. a. Given   **b.** Opposite sides of $\square$ are $\cong$.
c. Opposite sides of $\square$ are $\cong$.   **d.** Opposite $\angle$s of $\square$ are $\cong$.   **e.** SAS Congruence Postulate

Practice B

1. No, only one pair of opposite sides are parallel.   **2.** yes   **3.** No; it's a hexagon.   **4.** 13

5. 10.15   **6.** 6   **7.** 12   **8.** 5   **9.** 15

10. 68.3°   **11.** 50.7°   **12.** 61°   **13.** 68.3°

14. 46.3   **15.** $x = 9, y = 11$

16. $x = 30, y = 27, z = 36$ **17.** $x = 7, y = 3$

18.

Statements	Reasons
1. $\square ABCD$	1. Given
2. $\overline{AD} \cong \overline{BC}$	2. Opposite sides of $\square$ are $\cong$.
3. $\overline{AE} \cong \overline{CE}$, $\overline{DE} \cong \overline{BE}$	3. Diagonals of $\square$ bisect each other.
4. $\triangle AED \cong \triangle CEB$	4. SSS Congruence Postulate

19.

Statements	Reasons
1. $\square WXYZ$	1. Given
2. $\overline{ZM} \perp \overline{WX}$, $\overline{XN} \perp \overline{ZY}$	2. Given
3. $\angle ZMW, \angle XNY$ are right $\angle$'s	3. Definition of $\perp$
4. $\angle ZMW \cong \angle XNY$	4. All right $\angle$s are $\cong$.
5. $\angle W \cong \angle Y$	5. Opposite $\angle$s of $\square$ are $\cong$.
6. $\overline{ZW} \cong \overline{XY}$	6. Opposite sides of $\square$ are $\cong$.
7. $\triangle ZMW \cong \triangle XNY$	7. AAS Congruence Theorem

Practice C

1. No; consecutive $\angle$s are not supplementary.

2. No; opposite $\angle$s are not $\cong$.   **3.** yes   **4.** 3

5. 5   **6.** 4   **7.** 8   **8.** 5   **9.** 12   **10.** 37°

11. 90°   **12.** 53°   **13.** 53°   **14.** 20

15. $MP = 8\sqrt{2}, NO = 8\sqrt{2} \Rightarrow \overline{MP} \cong \overline{NO}$

16. $MN = 4, PO = 4 \Rightarrow \overline{MN} \cong \overline{PO}$

Lesson 6.2 *continued*

17. slope of $\overline{MP}$ = slope of $\overline{NO}$ = 1

18. yes; parallel lines have equal slope.

19. yes; $MQ = 2\sqrt{5}$, $QO = 2\sqrt{5}$, $\overline{MQ} \cong \overline{QO}$
$NQ = 2\sqrt{13}$, $QP = 2\sqrt{13}$, $\overline{NQ} \cong \overline{QP}$

20.

Statements	Reasons
1. $\square MATH$	1. Given
2. $\overline{MN} \cong \overline{AT}$	2. Given
3. $\overline{AT} \cong \overline{MH}$	3. Opposite sides of $\square$ are $\cong$.
4. $\overline{MN} \cong \overline{MH}$	4. Transitive Property of $\cong$
5. $\angle 1 \cong \angle 2$	5. $\angle$s opposite $\cong$ sides are $\cong$.

21.

Statements	Reasons
1. $\square ATRO$	1. Given
2. $\overline{PT} \cong \overline{IP}$	2. Given
3. $\angle I \cong \angle T$	3. $\angle$s opposite $\cong$ sides are $\cong$.
4. $\angle T \cong \angle AOR$	4. Opposite $\angle$s of a $\square$ are $\cong$.
5. $\angle I \cong \angle AOR$	5. Transitive Property of $\cong$

Reteaching with Practice

1. $a = 15$, $b = 135$ 2. $d = 10$, $e = 90$
3. $x = 24$, $y = 120$ 4. $a = 5$, $b = 3$, $c = 4$
5. $d = 98$, $e = 98$, $f = 82$
6. $g = 12$, $h = 9$, $i = 16$, $j = 14$
7. $x = 30$, $y = 6$ 8. $x = 40$, $y = 8$
9. $x = 10$, $y = 9$

Real-Life Application

1. 5 cm, 5 cm; Opposite sides of a parallelogram are congruent. 2. 110°; Opposite angles of a parallelogram are congruent.

3. 70°; Consecutive angles of a parallelogram are supplementary. 4. No. All four angles must be 90°. 5. Answers will vary.

Challenge: Skills and Applications

1. 5 2. $-2 < x < 3$ 3. 9
4. $x = 25$, $y = 10$ 5. $x = \pm 7$, $y = 10$

6. *Sample answer:*

Statements	Reasons
1. $IJKL$ is a parallelogram.	1. Given
2. $\overline{IJ} \parallel \overline{LK}$	2. Def. of parallelogram
3. $\angle JIK \cong \angle LKI$	3. Alternate Interior Angles Theorem
4. $\overline{LJ}$ bisects $\overline{IK}$.	4. Diagonals of a parallelogram bisect each other.
5. $\overline{IO} \cong \overline{KO}$	5. Def. of segment bisector
6. $\angle MOI \cong \angle NOK$	6. Vertical Angles Theorem
7. $\triangle MOI \cong \triangle NOK$	7. ASA Congruence Postulate
8. $\overline{MO} \cong \overline{NO}$	8. Corresp. parts of $\cong$ triangles are $\cong$.
9. O is the midpoint of $\overline{MN}$.	9. Def. of midpoint

7. *Sample answer:*

Statements	Reasons
1. $ABCD$ and $EFGH$ are parallelograms.	1. Given
2. $\overline{AB} \parallel \overline{CD}$	2. Def. of parallelogram
3. $\angle FAE \cong \angle EHD$	3. Alternate Interior Angles Theorem
4. $\overline{EH} \parallel \overline{FG}$	4. Def. of parallelogram
5. $\angle EHD \cong \angle HCG$	5. Corresp. Angles Postulate
6. $\angle FAE \cong \angle HCG$	6. Transitive Prop. of Congruence
7. $\angle HEF \cong \angle FGH$	7. Opposite angles of a parallelogram are congruent.
8. $\angle HEF$ and $\angle AEF$ are supplementary; $\angle FGH$ and $\angle CGH$ are supplementary.	8. Linear Pair Post.
9. $\angle AEF \cong \angle CGH$	9. Congruent Supplements Thm.
10. $\overline{EF} \cong \overline{HG}$	10. Opposite sides of a parallelogram are congruent.
11. $\triangle FAE \cong \triangle HCG$	11. AAS Congruence Theorem

Lesson 6.2 *continued*

8. Check constructions. Students will need to use the construction they have learned for parallel lines two times. To draw the missing short side of the parallelogram, they construct a line through the unattached endpoint of the long side that is parallel to the existing short side. To draw the missing long side of the parallelogram, they construct a line through the unattached endpoint of the short side that is parallel to the existing long side.

Lesson 6.3

Warm-Up Exercises

1. SSS Cong. Post.   **2.** Opposite sides of a $\square$ are $\cong$.   **3.** Diagonals of a $\square$ bisect each other.

Daily Homework Quiz

1. 24; the diags. of a $\square$ bisect each other, so $KH = 5$ and $KG = 7$. Then $FH = 5 + 5 = 10$ and $JG = 7 + 7 = 14$, so $FH + JG = 10 + 14 = 24$.

2. They are $\cong$. Opp. sides of a $\square$ are $\cong$, so $\overline{AB} \cong \overline{CH}$, $\overline{CH} \cong \overline{DG}$, and $\overline{DG} \cong \overline{EF}$. Using the Transitive Prop. of Cong. twice, $\overline{AB} \cong \overline{EF}$.

Lesson Opener

Allow 15 minutes.

1. yes   **2.** yes   **3.** yes

4. no; *Sample counterexample:*

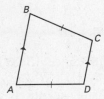

5. yes   **6.** yes

7. no; *Sample counterexample:*

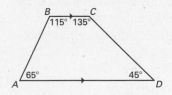

8. yes

9. no; *Sample counterexample:*

10. yes

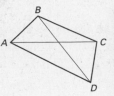

Practice A

1. yes; 1 pair of opposite sides are $\parallel$ and $\cong$.

2. yes; diagonals bisect each other.

3. yes; both pairs of opposite sides are $\cong$.

4. no; need same pair of opposite sides $\parallel$ and $\cong$.

5. yes; could show both pairs of opposite sides $\parallel$.

6. yes; consecutive $\angle$ are supplementary.

7. *Sample answer:* $\overline{AD} \parallel \overline{BC}$

8. *Sample answer:* $\overline{AD} \cong \overline{BC}$

9. *Sample answer:* $\overline{AB} \cong \overline{CD}$

10. *Sample answer:* $\overline{AE} \cong \overline{CE}$

11. *Sample answer:*
$m\angle DAB + m\angle ABC = 180°$

12. $x = 2, y = 5$ **13.** $x = 35, y = 25$

14. $x = 4, y = 1$

15. a.

Statements	Reasons
1. $\triangle MJK \cong \triangle KLM$	**1.** Given
2. $\overline{JK} \cong \overline{LM}$ $\overline{JM} \cong \overline{LK}$	**2.** Corresp. parts of $\cong \triangle$'s are $\cong$.
3. $MJKL$ is a $\square$.	**3.** If both pairs of opp. sides of quad. are $\cong$, then quad is a $\square$.

b.

Statements	Reasons
1. $\triangle MJK \cong \triangle KLM$	**1.** Given
2. $\overline{JK} \cong \overline{LM}$ $\angle JKM \cong \angle KML$	**2.** Corresp. parts of $\cong \triangle$'s are $\cong$.
3. $\overline{JK} \parallel \overline{LM}$	**3.** Alternate Interior $\angle$'s Converse
4. $MJKL$ is a $\square$.	**4.** If one pair of opp. sides are $\cong$ and $\parallel$, then quad is a $\square$.

Practice B

1. yes   **2.** no   **3.** no   **4.** yes   **5.** no   **6.** yes

7. $x = 6, y = 4$ **8.** $x = 5, y = 75$

9. $x = 12, y = 7$

Answers

Lesson 6.3 *continued*

10. *Sample answer:* slope of $\overline{AB}$ = slope of $\overline{CD}$ = −4 and slope of $\overline{BC}$ = slope of $\overline{AD}$ = $\frac{2}{5}$; so *ABCD* is a ▱ by definition.

11. *Sample answer:* $AB = CD = \sqrt{45}$ and $BC = DA = \sqrt{65}$ so *ABCD* is a ▱ since both pairs of opposite sides are ≅.

12.

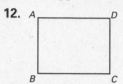

13. Yes, $\overline{AB} \parallel \overline{DC}$ and $\overline{AD} \parallel \overline{BC}$

14.

Statements	Reasons
1. $\angle AFD \cong \angle ADF$	1. Given
2. $\overline{AD} \cong \overline{AF}$	2. Sides opp. ≅ ⦞ are ≅.
3. $\overline{AF} \cong \overline{BC}$	3. Given
4. $\overline{AD} \cong \overline{BC}$	4. Transitive Prop. of ≅
5. $\overline{AB} \cong \overline{CD}$	5. Given
6. *ABCD* is a ▱.	6. If both pairs of opp. sides are ≅, then quad. is a ▱.

15.

Statements	Reasons
1. $\triangle RQP \cong \triangle ONP$ *R* is midpoint of $\overline{MQ}$.	1. Given
2. $\overline{MR} \cong \overline{RQ}$	2. Definition of midpoint
3. $\overline{RQ} \cong \overline{NO}$	3. Corresp. parts of ≅ △'s are ≅.
4. $\overline{MR} \cong \overline{NO}$	4. Transitive Prop. of ≅
5. $\angle QRP \cong \angle NOP$	5. Corresp. parts of ≅ △'s are ≅.
6. $\overleftrightarrow{MQ} \parallel \overleftrightarrow{NO}$	6. Alternate Interior ∠'s Converse
7. $\overline{MR} \parallel \overline{NO}$	7. If two lines are ∥, segments combined within them are ∥
8. *MRON* is a ▱	8. If one pair of opp. sides are ∥ and ≅, then quad. is a ▱.

Practice C

1. yes **2.** yes **3.** no **4.** no **5.** no
6. yes **7.** yes **8.** yes

9. slope of $\overline{AB}$ = slope of $\overline{CD}$ = −1 and slope of $\overline{BC}$ = slope of $\overline{DA}$ = 5, so *ABCD* is a ▱ by definition

10. $AB = CD = \sqrt{17}$ and $BC = DA = \sqrt{45}$ so *ABCD* is a ▱ since both pair of opposite sides are ≅. **11.** $(8, 6), (0, -8), (-8, 10)$

12. $(6, -1), (0, -7), (-4, 5)$

13.

Statements	Reasons
1. Regular hexagon *JKLMNO*	1. Given
2. $\overline{JO} \cong \overline{NM}$, $\overline{JK} \cong \overline{ML}$, $\angle J \cong \angle M$	2. Definition of regular polygon
3. $\triangle OJK \cong \triangle NML$	3. SAS Congruence Postulate
4. $\overline{OK} \cong \overline{NL}$	4. Corresp. parts of ≅ △'s are ≅.
5. $\overline{ON} \cong \overline{KL}$	5. Definition of regular polygon
6. *OKLN* is a ▱	6. If both pairs of opp. sides are ≅, then quad. is a ▱.

14.

Statements	Reasons
1. *VWKJ* and *SJRU* are ▱.	1. Given
2. $\angle W \cong \angle J$, $\angle J \cong \angle U$	2. Opp. ⦞ of a ▱ are ≅.
3. $\angle W \cong \angle U$	3. Transitive Prop. of ≅

15. Since *ABCD* is a ▱, opposite sides are ∥. So $\overline{AD} \parallel \overline{BC}$ and segments contained within ∥ segments are also ∥ so $\overline{AE} \parallel \overline{BF}$. We know opposite sides of ▱ are ≅ so $\overline{AD} \cong \overline{BC}$. Since *E* and *F* are given as midpoints we can show $\overline{AE} \cong \overline{ED}$ and $\overline{CF} \cong \overline{FB}$, so through Segment Addition Postulate and Division Property of Equality we can show $\overline{AE} \cong \overline{FB}$. So quad. *ABFE* is a ▱ since one pair of opposite sides are ∥ and ≅.

Reteaching with Practice

1. Slope of $\overline{AB}$ = slope of $\overline{CD}$ = −4; slope of $\overline{AD}$ = slope of $\overline{BC}$ = $\frac{2}{5}$

2. $AB = CD = \sqrt{26}$; $AD = BC = \sqrt{45} = 3\sqrt{5}$

Lesson 6.3 *continued*

3. *Sample answer:* Slope of $\overline{AB}$ = slope of $\overline{CD} = \frac{1}{4}$; $AB = CD = \sqrt{17}$

4. *Sample answer:* Slope of $\overline{AB}$ = slope of $\overline{CD} = -\frac{1}{6}$; slope of $\overline{AD}$ = slope of $\overline{BC} = 9$

Interdisciplinary Application

1. 1. Given   2. If the diagonals of a quadrilateral bisect each other, then the quadrilateral is a parallelogram.

2. *Sample answer:*

Statements	Reasons
1. $\overline{AE} \cong \overline{EC}, \overline{BE} \cong \overline{ED}$	1. Given
2. $\angle AEB \cong \angle CED$, $\angle BEC \cong \angle AED$	2. Verticle Angles Theorem
3. $\triangle AEB \cong \triangle CED$, $\triangle BEC \cong \triangle AED$	3. SAS Congruence Postulate
4. $\overline{AB} \cong \overline{CD}, \overline{BC} \cong \overline{AD}$	4. Corresponding parts of congruent triangles are congruent.
5. $ABCD$ is a parallelogram.	5. If both pairs of opposite sides of a quadrilateral are congruent, then the quadrilateral is a parallelogram.

Math & History Application

1. The octagon's area is approximately equal to the circle's area, and the octagon's area can be found by subtracting the unused area of the four triangles in each corner from the area of the entire 3-by-3 grid; 7 square units

2.
$$A = \pi r^2$$
$$7 \approx \pi \left(\frac{3}{2}\right)^2$$
$$7 \approx \pi \left(\frac{9}{4}\right)$$
$$\frac{28}{9} \approx \pi$$
$$3.111\ldots \approx \pi$$

Challenge: Skills and Applications

1. *Sample answer:*

Statements	Reasons
1. $UWXZ$ is a parallelogram.	1. Given
2. $\angle Z \cong \angle W$	2. Opposite angles of a $\square$ are $\cong$.
3. $\angle 1 \cong \angle 8$	3. Given
4. $\angle 5 \cong \angle 4$	4. Third Angles Thm.
5. $\overline{UW} \parallel \overline{ZX}$	5. Def. of $\square$
6. $\angle 4 \cong \angle 7$	6. Alternate Interior Angles Theorem
7. $\angle 5 \cong \angle 7$	7. Transitive Prop. of Congruence
8. $\overline{UY} \parallel \overline{VX}$	8. Corresp. Angles Converse
9. $\overline{UV} \parallel \overline{YX}$	9. This follows from Step 5.
10. $UVXY$ is a parallelogram.	10. Def. of $\square$

2. *Sample answer:*

Statements	Reasons
1. $GIJL$ is a parallelogram.	1. Given
2. $\overline{GI} \parallel \overline{LJ}$	2. Def. of $\square$
3. $\angle GIL \cong \angle JLI$	3. Alternate Interior Angles Theorem
4. $\overline{GJ}$ bisects $\overline{LI}$	4. Diagonals of a $\square$ bisect each other.
5. $\overline{MI} \cong \overline{ML}$	5. Def. of segment bisector
6. $\angle HMI \cong \angle KML$	6. Vertical $\angle$s Thm.
7. $\triangle HMI \cong \triangle KML$	7. ASA Congruence Postulate
8. $\overline{MH} \cong \overline{MK}$	8. Corresp. parts of $\cong$ $\triangle$s are $\cong$.
9. $\overline{HK}$ and $\overline{IL}$ bisect each other.	9. Def. of segment bisector
10. $HIKL$ is a parallelogram.	10. If the diagonals of a quad. bisect each other, then it is a $\square$.

Lesson 6.3 *continued*

3. *Sample answer:* The conjecture cannot be proved, because it is not true. (See Exercise 5.) A student may attempt to prove the conjecture by drawing quadrilateral $ABCD$ with $\overline{AB} \cong \overline{CD}$ and $\angle B \cong \angle D$, and then drawing diagonal $\overline{AC}$ to see if $\triangle ABC$ and $\triangle CDA$ must be congruent. Since $\overline{AC} \cong \overline{CA}$, two sides and an angle of $\triangle ABC$ are congruent to two sides and an angle of $\triangle CDA$. However, you cannot conclude that the triangles are congruent since the angle that is congruent is **not** the included angle. (That is, there is no SSA Congruence Postulate.)

4.

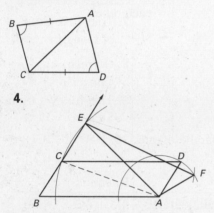

5. a. *Sample answer:* By construction, $\overline{AC} \cong \overline{AE}$, $\overline{AD} \cong \overline{AF}$, and $m\angle CAD = m\angle EAF$. Therefore, by the SAS Congruence Postulate, $\triangle ACD \cong \triangle AEF$. Since corresponding parts of congruent triangles are congruent, $\overline{DC} \cong \overline{FE}$ and $\angle ADC \cong \angle AFE$. Since $ABCD$ is a parallelogram, $\overline{AB} \cong \overline{DC}$ and $\angle ABE \cong \angle ADC$. Therefore, by the Transitive Property of Congruence, $\overline{AB} \cong \overline{FE}$ and $\angle ABE \cong \angle AFE$.   **b.** false; *Sample answer:* $ABEF$ is a quadrilateral with a pair of congruent opposite sides and a pair of congruent opposite angles, but it is not a parallelogram.

Quiz 1

1. convex, equilateral   **2.** 20; the sum of the measures of the interior $\angle$s of a quad. is 360°, so $4y + 4y + 100 + 100 = 360$, $8y = 160$, $y = 20$.

3.

Statements	Reasons
1. $m\angle 1 = 90°$, $\overline{AB} \parallel \overline{CE} \parallel \overline{GF}$	1. Given
2. $\angle 1 \cong \angle 2$	2. Vertical angles are congruent.
3. $m\angle 1 + m\angle B = 180°$, $m\angle 2 + m\angle G = 180°$ $m\angle C + m\angle 1 = 180°$ $m\angle 2 + m\angle E = 180°$	3. Consecutive Interior Angles Theorem
4. $m\angle B = m\angle C =$ $m\angle E = m\angle G = 90°$	4. Subtraction property of equality
5. $m\angle A = m\angle F = 90°$	5. Interior Angles of a Quad. Theorem
6. $\angle A$ and $\angle F$ are rt $\angle$s	6. Definition of right angle

4. *Sample answers:* Use slopes to show that both pairs of opp. sides are $\parallel$, use the Distance Formula to show that both pairs of opp. sides are $\cong$, use slope and the Distance Formula to show that one pair of opp. sides are both $\parallel$ and $\cong$, use the Midpoint Formula to show that the diagonals bisect each other.

Lesson 6.4

Warm-Up Exercises

1. 16   **2.** $\sqrt{5}$ **3.** 2   **4.** yes

Daily Homework Quiz

1. Since corresp. parts of $\cong$ $\triangle$'s are $\cong$, $\angle C \cong \angle G$ and $\angle E \cong \angle A$. Both pairs of opp. $\angle$s of $ACEG$ are $\cong$, so $ACED$ is a $\square$.   **2.** Using $\overline{EF}$ and $\overline{GH}$: $EF = GH = \sqrt{85}$, so $\overline{EF} \cong \overline{GH}$. Slope of $\overline{EF}$ = slope of $\overline{GH} = \frac{7}{6}$, so $\overline{EF} \parallel \overline{GH}$.

3. The midpoints of diags. $\overline{JL}$ and $\overline{KM}$ are both $(-1.5, 1.5)$, so they bisect each other.

Lesson Opener

Allow 15 minutes.

1. yes   **2.** yes; yes; 90°   **3.** yes   **4.** angle measures   **5. a.** angle measures   **b.** side lengths   **c.** angle measures   **d.** side lengths   **e.** both   **f.** both

Lesson 6.4 *continued*

Technology Activity

1. Check student's work. **2.** Check student's work. **3.** Check student's work.

Practice A

1. rhombus; diagonals are ⊥ but not ≅.

2. square; adjacent sides are ≅ and ∠s are right angles. **3.** rectangle; angles are right angles.

4. rhombus; adjacent sides are ≅.

5. rectangle; diagonals are ≅.

6. square; diagonals are ≅ and ⊥.

7. B, D **8.** A, B, C, D **9.** A, B, C, D

10. B, D **11.** C, D **12.** D **13.** rhombus

14. rectangle **15.** square **16.** parallelogram

17. 11 **18.** 13 **19.** 5

Practice B

1. always **2.** sometimes **3.** sometimes

4. always **5.** always **6.** always **7.** 32°

8. 86° **9.** 66° **10.** 35° **11.** 16 **12.** 26

13. *Sample answer:* Rectangle; adjacent sides are ⊥ but not ≅.

14. *Sample answer:* Rhombus; adjacent sides are ≅ but not ⊥.

15. *Sample answer:* Rectangle; diagonals are ≅ but not ⊥.

16. *Sample answer:* Square; diagonals are ≅ and ⊥.

17.

Statements	Reasons
1. ▱HIJK	**1.** Given
2. △HOI ≅ △JOI	**2.** Given
3. $\overline{IH} \cong \overline{IJ}$	**3.** Corresp. parts of ≅ △'s are ≅.
4. HIJK is a rhombus	**4.** A ▱ with adjacent sides ≅ is a rhombus.

18.

Statements	Reasons
1. Rectangle RECT	**1.** Given
2. $\overline{RT} \cong \overline{EC}$, $\overline{RT} \parallel \overline{EC}$	**2.** Property of Rectangle
3. ∠ART ≅ ∠ACE	**3.** Alt. Int. ∠s are ≅.
4. ∠RAT ≅ ∠CAE	**4.** Vertical ∠s are ≅.
5. △ART ≅ △ACE	**5.** AAS Congruence Thm.

Practice C

1. 27° **2.** 54° **3.** 126° **4.** 54° **5.** 126°

6. 90° **7.** 90° **8.** 63° **9.** true; false

10. false; true **11.** true; false **12.** false; false

13. true; true; a rhombus is a square if and only if it is a rectangle. **14.** 11 **15.** 60° **16.** 10

17.

Statements	Reasons
1. WHAT is a ▱.	**1.** Given
2. $\overline{WD} \cong \overline{DA}$	**2.** Diagonals of ▱ bisect each other.
3. $\overline{HD} \cong \overline{DT}$	**3.** Diagonals of ▱ bisect each other.
4. DART is a rhombus.	**4.** Given
5. $\overline{DT} \cong \overline{DA}$	**5.** Definition of rhombus
6. $\overline{WD} \cong \overline{HD} \cong \overline{DA} \cong \overline{DT}$	**6.** Substitution
7. WA = WD + DA, HT = HD + DT	**7.** Seg. Add. Postulate
8. WA = HT	**8.** Substitution
9. WHAT is a rectangle	**9.** If diagonals of ▱ are ≅, then it is a rectangle.

18.

Statements	Reasons
1. △GEC ≅ △GHX	**1.** Given
2. $\overline{GE} \cong \overline{GH}$	**2.** Corresp. parts of ≅ △'s are ≅.
3. GEBH is a ▱	**3.** Given
4. $\overline{GH} \cong \overline{EB}$	**4.** If quad. is a ▱, then opp. sides are ≅.
5. $\overline{GE} \cong \overline{HB}$	**5.** If quad. is a ▱, then opp. sides are ≅.
6. $\overline{GE} \cong \overline{EB}$	**6.** Substitution
7. $\overline{EB} \cong \overline{HB}$	**7.** Substitution
8. GEBH is a rhombus.	**8.** A ▱ with 4 ≅ sides is a rhombus.

Lesson 6.4 *continued*

19.

Statements	Reasons
1. *JXPE* is a ▱.	1. Given
2. $\overline{EJ} \cong \overline{PX}$, $\overline{EP} \cong \overline{JX}$	2. Opp. sides of ▱ are ≅.
3. $\overline{EX} \cong \overline{EX}$	3. Reflex. Prop. of ≅
4. △*JXE* ≅ △*PEX*	4. SSS Congr. Postulate
5. ∠*J* ≅ ∠*XPE*	5. Corresp. parts of ≅ △'s are ≅.
6. $\overline{XP} \perp \overline{EN}$	6. Given
7. ∠*XPE* is a rt. ∠.	7. Definition of ⊥
8. ∠*J* is a rt. ∠.	8. Substitution
9. *JANE* is a ▱.	9. Given
10. ∠*J* and ∠*A* are supplementary.	10. Adj. ∡ of ▱ are supplementary.
11. ∠*A* is a rt. ∠.	11. If 2 ∡ are suppl. and one is a rt. ∠, then the other is a rt. ∠.
12. ∠*N* and ∠*PEJ* are rt. ∡.	12. Opp. ∡ of a ▱ are ≅.
13. *JANE* is a rectangle.	13. A ▱ with 4 rt. ∡ is a rectangle.

Reteaching with Practice

1. 4 ≅ sides, opposite sides are ∥ and ≅, opposite ∡ are ≅, consec. ∡ are supplementary, diagonals bisect each other, diagonals are ⊥, diagonals bisect opposite ∡

2. opposite sides are ∥, sides meet at right ∡, both pairs of opposite sides are ≅, opposite ∡ are ≅, consec. ∡ are supplementary, diagonals bisect each other, diagonals are ≅

3. always

4. sometimes
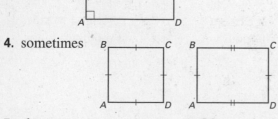

5. always
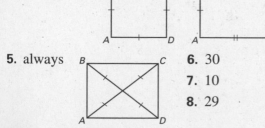

6. 30
7. 10
8. 29

Real-Life Application

1. a. rhombus **b.** rhombus, rectangle, square
c. rectangle

2.

Shape	Rhombus	Rectangle	Square
Number	20	46	16

Challenge: Skills and Applications

1. $x = 26$, $y = -3$ or $y = 8$ **2.** $x = 4$, $y = \pm 3$
3. $x = -3$ or $x = 5$, $y = 4$ **4.** $x = 9$, $y = \pm 3$
5. $x = 7$, $y = 35$
6. *Sample answer:*

Statements	Reasons
1. $\overline{BD}$ is a ⊥ bisector of $\overline{AC}$.	1. Given
2. $\overline{BD} \perp \overline{AC}$	2. Def. of ⊥ bisector
3. ∠*AEB* and ∠*CED* are right angles.	3. If 2 lines are ⊥, then they form 4 right angles.
4. ∠*AEB* ≅ ∠*CED*	4. Right Angle Congruence Thm.
5. $\overline{AE} \cong \overline{CE}$	5. Def. of ⊥ bisector
6. $\overline{AD} \parallel \overline{BC}$	6. Given
7. ∠*BAE* ≅ ∠*DCE*	7. Alternate Interior Angles Thm.
8. △*AEB* ≅ △*CED*	8. ASA Congruence Postulate
9. $\overline{AB} \cong \overline{CD}$	9. Corresp. parts of ≅ △s are ≅.
10. $\overline{AB} \cong \overline{BC}$, $\overline{CD} \cong \overline{DA}$	10. ⊥ Bisector Theorem
11. $\overline{AB}$, $\overline{BC}$, $\overline{CD}$, and $\overline{DA}$ are congruent.	11. Transitive Prop. of Congruence
12. *ABCD* is a rhombus.	12. Rhombus Corollary

7. *Sample answer:* Since *IJKL* is a parallelogram, $\overline{IK}$ and $\overline{JL}$ bisect each other. So, *M* is the midpoint of $\overline{IK}$ and of $\overline{JL}$. Therefore, $IK = 2IM$ and $JL = 2JM$. But ∠1 ≅ ∠2 (given), so by the Base Angles Converse, $IM = JM$. Therefore, by properties of equality, $2IM = 2JM$ and $IK = JL$. Therefore, *IJKL* is a parallelogram with congruent diagonals, so *IJKL* is a rectangle.

Lesson 6.4 *continued*

8. *Sample answer:*

Statements	Reasons
1. *WXYZ* is a parallelogram.	1. Given
2. $\overline{WX} \parallel \overline{ZY}$	2. Def. of parallelogram
3. $\angle 2 \cong \angle 3$	3. Alternate Interior Angles Theorem
4. $\angle 1 \cong \angle 2$	4. Given
5. $\angle 1 \cong \angle 3$	5. Transitive Prop. of Congruence
6. $\overline{YZ} \cong \overline{ZW}$	6. Base Angles Converse
7. $\overline{WX} \cong \overline{YZ}, \overline{XY} \cong \overline{ZW}$	7. Opposite sides of a parallelogram are congruent.
8. $\overline{WX}, \overline{XY}, \overline{YZ},$ and $\overline{ZW}$ are congruent.	8. Transitive Prop. of Congruence
9. *WXYZ* is a rhombus.	9. Rhombus Corollary

9. no; *Sample answer:* The construction worker also needs to verify that the shape of the opening is a parallelogram.

Lesson 6.5

Warm-Up Exercises

1. -1 **2.** 1 **3.** $2\sqrt{41}$ **4.** 12

Daily Homework Quiz

1. always **2.** Sometimes; this is true if *ABCD* is also a rectangle, which means that it must be a square. **3.** always **4.** square **5.** 1

Lesson Opener

Allow 5 minutes.

1. 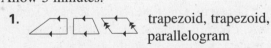 trapezoid, trapezoid, parallelogram

2.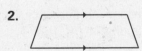

The nonparallel sides are congruent. Diagonals are congruent; base angles are congruent.

Practice A

1. A **2.** B **3.** C **4.** E **5.** D
6. never **7.** always **8.** always
9. sometimes **10.** $m\angle B = 53°$, $m\angle C = m\angle D = 127°$ **11.** $m\angle C = 48°$, $m\angle D = 89°$ **12.** $m\angle B = 108°$, $m\angle C = m\angle D = 72°$ **13.** 10 **14.** 10.5
15. 20 **16.** $KI = KE \approx 4.47, IT = ET \approx 6.32$
17. $m\angle I = 124°, m\angle K = 80°$
18. $m\angle E = 118°, m\angle K = 50°$

Practice B

1. E **2.** F **3.** B **4.** C **5.** D **6.** A
7. $m\angle D = 112°, m\angle B = m\angle A = 68°$
8. $m\angle A = 90°, m\angle C = 48°$ **9.** $m\angle D = 77°$, $m\angle A = m\angle B = 103°$ **10.** 10.5 **11.** 14
12. 2 **13.** $AM = MH \approx 7.81, AT = HT \approx 9.43$
14. $m\angle A = m\angle H = 129.5°$
15. $m\angle A = 118°, m\angle T = 43°$
16.

Statements	Reasons
1. $\overline{DE} \parallel \overline{AV}$	1. Given
2. *DAVE* is a trapezoid.	2. Def. of trapezoid
3. $\triangle DAV \cong \triangle EVA$	3. Given
4. $\overline{DA} \cong \overline{EV}$	4. Corresp. parts of $\cong$ $\triangle$'s are $\cong$.
5. *DAVE* is an isos. trap.	5. Def. of isos. trap.

17.

Statements	Reasons
1. $\overline{WV}$ is midsegment of $\triangle XYZ$.	1. Given
2. $\overline{WV} \parallel \overline{XY}$	2. Midsegment Thm.
3. $\overline{XZ} \cong \overline{ZY}$	3. Given
4. $\angle X \cong \angle Y$	4. Base $\angle$s Theorem
5. *XWVY* is an isos. trap.	5. If trap. has $\cong$ base $\angle$s then it is an isos. trap.

Practice C

1. yes, no **2.** no **3.** yes, yes
4. 107° **5.** 129° **6.** 68° **7.** 20.5 **8.** 20
9. 7 **10.** $WE = SE \approx 7.62; WT = ST \approx 10.44$

Lesson 6.5 *continued*

11. $m\angle E = m\angle T = 118°$ **12.** $m\angle T = 113°$, $m\angle W = 97°$ **13.** 60°, 60°, 120°, 120°

14. The sum of the lengths of the two bases is 12.

15.

Statements	Reasons
1. *LORI* is a rectangle.	**1.** Given
2. $\angle ILB$ and $\angle ROD$ are rt. &.	**2.** Def. of rectangle
3. $\angle ILB \cong \angle ROD$	**3.** All right & are ≅.
4. $\overline{LI} \cong \overline{OR}$	**4.** Opp. sides of ▱ are ≅.
5. $\overline{LB} \cong \overline{DO}$	**5.** Given
6. $\triangle LBI \cong \triangle ODR$	**6.** SAS Congruence Postulate
7. $\overline{BI} \cong \overline{DR}$	**7.** Corresp. parts of ≅ △'s are ≅.
8. $\overline{BD} \parallel \overline{IR}$	**8.** Def. of ▱
9. *BIRD* is an isos. trap.	**9.** Def. of isos. trap.

16.

Statements	Reasons
1. $\overline{AF} \not\cong \overline{BC}$	**1.** Given
2. $\triangle ABC \cong \triangle CDA$	**2.** Given
3. $\angle CAB \cong \angle ACD$	**3.** Corresp. parts of ≅ △'s are ≅.
4. $\overline{CF} \parallel \overline{AB}$	**4.** Alternate Interior & Thm. Converse
5. *ABCF* is a trapezoid.	**5.** Def. of trapezoid

Reteaching with Practice

1. 7 **2.** 9 **3.** 118 **4.** 110 **5.** 35 **6.** 5

Interdisciplinary Application

1. a. isosceles trapezoid; *ABDC* is an isosceles trapezoid because it has exactly one pair of parallel sides and a pair of congruent base angles.
b. trapezoid; *GHFD* is a trapezoid because it has exactly one pair of parallel sides.

2. *Sample answer:*

Statements	Reasons
1. *ABFE* is a trapezoid.	**1.** Given
2. $\overline{CD} \parallel \overline{EF}$, $\angle ACD \cong \angle BDC$	**2.** Given
3. $\angle ACD \cong \angle AEF$, $\angle BDC \cong \angle BFE$	**3.** Corresponding Angles Postulate
4. $\angle BDC \cong \angle AEF$	**4.** Substitution
5. $\angle AEF \cong \angle BFE$	**5.** Transitive Property of Congruence
6. *ABFE* is an isosceles trapezoid.	**6.** If a trapezoid has a pair of congruent base angles, then it is an isosceles trapezoid.

3. 2.5 cm **4.** 0.6 cm

Challenge: Skills and Applications

1. 3, 6 **2.** 4 **3.** 8 **4.** *Sample answer:* Draw line *k* through *B*, parallel to $\overline{AD}$ (Parallel Postulate). Line *k* is in the same plane as *ABCD* and is not parallel to $\overleftrightarrow{DC}$, so we may let *E* be the point where line *k* intersects $\overline{DC}$. Now $\angle C \cong \angle D$ (given), and $\angle D \cong \angle BEC$ (Corresponding Angles Postulate), so $\angle C \cong \angle BEC$. Now $\overline{BC} \cong \overline{BE}$ (Base Angles Converse) and $\overline{BE} \cong \overline{AD}$ (because *ABED* is a parallelogram), so $\overline{BC} \cong \overline{AD}$ (Transitive property of equality). So, *ABCD* is an isosceles trapezoid.

5. *Sample answer:* We are given $\overline{PQ} \cong \overline{PS}$, so by the Converse of the Perpendicular Bisector Theorem, *P* is on a perpendicular bisector of $\overline{QS}$. Therefore, since $\overline{PR} \perp \overline{QS}$ (given), $\overline{PR}$ is a perpendicular bisector of $\overline{QS}$. (This is a somewhat subtle application of the Perpendicular Postulate. Since $\overleftrightarrow{PR}$ is perpendicular to $\overline{QS}$ and contains *P*, it must be the same line as the perpendicular bisector that contains *P*.) Therefore, by the Perpendicular Bisector Theorem, $\overline{QR} \cong \overline{SR}$. Since $\overline{PQ} \cong \overline{PS}$ and $\overline{QR} \cong \overline{SR}$, *PQRS* is either a kite or a rhombus. If *PQRS* were a rhombus, it would also be a parallelogram, so its diagonals would bisect each other. Since *T* is not the midpoint of $\overline{PR}$, *PQRS* cannot be a rhombus. Therefore, *PQRS* is a kite.

6. $(0, -2)$, $(6, 0)$, $\left(\frac{6}{5}, \frac{8}{5}\right)$ **7.** $(-10, 15)$

8. $(-6, 8)$, $(6, 8)$

Lesson 6.5 *continued*

Quiz 2

1. rhombus   **2.** rectangle   **3.** trapezoid

4. kite   **5.** 4

6.

Statements	Reasons
1. *ABCD* is an isosceles trapezoid.	**1.** Given
2. $\angle A \cong \angle B$	**2.** Base $\angle$s of an isos. trap. are $\cong$.
3. $\overline{DE} \perp \overline{AB}$; $\overline{CF} \perp \overline{AB}$	**3.** Given
4. $\overline{DE} \parallel \overline{CF}$	**4.** Two lines $\perp$ to the same line are $\parallel$.
5. *DCFE* is a rectangle.	**5.** Opposite sides are $\parallel$ and all int $\angle$s are right $\angle$s.
6. $\overline{DE} \cong \overline{CF}$	**6.** Opp. sides of a rect. are $\cong$.
7. $\triangle ADE \cong \triangle BCF$	**7.** AAS Congruence Postulate
8. $\overline{AD} \cong \overline{CB}$	**8.** Corresponding parts of congruent triangles are $\cong$.

Lesson 6.6

Warm-Up Exercises

1. kite   **2.** square   **3.** trapezoid   **4.** rhombus

Daily Homework Quiz

1. $m\angle F = 74°$, $m\angle G = 74°$, $m\angle H = 106°$

2. 21   **3.** $ED = GD = 4.47$, $EF = FG = 7.21$

4. No; $AB = BC = 2\sqrt{5}$, but $CD = \sqrt{85} \neq \sqrt{73} = AD$, so only one pair of consec. sides is $\cong$.

Lesson Opener

Allow 10 minutes.

1–4. Answers will vary. Sample answers are given.   **1.** trapezoid; it is not a parallelogram.   **2.** isosceles trapezoid; its diagonals are not perpendicular.   **3.** rhombus; its sides are all congruent.   **4.** rectangle; its angles are all congruent.

Practice A

1. A, B, C, D   **2.** A, B, C, D   **3.** A, B, C, D

4. E, F   **5.** F   **6.** G   **7.** B, D, F   **8.** C, D, G

9. rectangle   **10.** isosceles trapezoid

11. square   **12–14.** Answers vary; *Sample answers:* **12.** $\overline{AC} \cong \overline{BD}$; $\square$ with $\cong$ diagonals is a rectangle. **13.** $\overline{AD} \cong \overline{BC}$; quad. with opp. sides $\cong$ is $\square$. **14.** $\overline{AD} \cong \overline{BC}$; a trap. with non-parallel sides $\cong$ is isosceles. **15.** Parallelogram; $PQ = SR = \sqrt{26}$ and $PS = QR = \sqrt{10}$

16. kite; $SP = SR = 3$ and $QP = QR = \sqrt{29}$

Practice B

$\square$	Rect.	Rhom.	Sq.	Trap.	Kite
1. X	X	X	X		
2.	X		X		
3.		X	X		X
4. X	X	X	X		
5. X	X	X	X		
6. X	X	X	X		

7. isosceles trapezoid   **8.** parallelogram, rectangle, rhombus, square   **9.** parallelogram, rhombus, rectangle, square   **10–12.** Answers vary; *Sample answers:* **10.** $\overline{AD} \cong \overline{DC}$; a $\square$ with adj. sides $\cong$ is a rhombus. **11.** $\overline{BC} \cong \overline{AD}$; Trap. with nonparallel sides $\cong$ is isosceles. **12.** $\overline{AD} \cong \overline{BC} \cong \overline{AB}$; a rhombus with rt. $\angle$s is a square.

13. parallelogram; $PQ = RS = \sqrt{104}$ and $QR = SP = 5$ **14.** rectangle; slope of $\overline{QP}$ = slope of $\overline{RS} = -\frac{1}{4}$; slope of $\overline{PS}$ = slope of $\overline{QR} = 4$; adjacent sides are $\perp$ (slopes are negative reciprocals) but not congruent $\left(\sqrt{17} \neq \sqrt{68}\right)$

15. rhombus; $PQ = QR = RS = SP = \sqrt{50}$; diagonals are not congruent ($14 \neq 2$)

16. (4.5, 5.5), (11.5, 5.5), (11.5, 4.5), (4.5, 4.5); rectangle

Practice C

1.

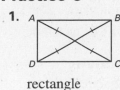

rectangle

2.

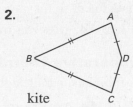

kite

3.

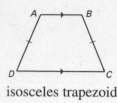

isosceles trapezoid

4.

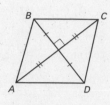

rhombus

5.

square

6. sometimes

7. always 8. always

9. never 10. sometimes

11. never 12. sometimes

13–15. Answers vary; *Sample answers:*

13. $\overline{AC} \cong \overline{BD}$; parallelogram, because $\overline{AD} \parallel \overline{BC}$ and $\overline{AD} \cong \overline{BC}$; parallelogram with $\cong$ diagonals is a rectangle.

14. $\overline{AB} \cong \overline{AD}$; quad. with two pairs of consec. $\cong$ sides, but opposite sides $\not\cong$, is a kite.

15. $\overline{AD} \cong \overline{BC}$; trap. with nonparallel sides $\cong$ is isosceles. **16.** parallelogram; slope of $\overline{PQ}$ = slope of $\overline{RS} = -\frac{1}{5}$; slope of $\overline{QR}$ = slope of $\overline{PS} = 1$; adjacent sides $\not\cong \left(\sqrt{18} \neq \sqrt{26}\right)$

17. rectangle; slope of $\overline{PQ}$ = slope of $\overline{RS} = -\frac{1}{4}$; slope of $\overline{QR}$ = slope of $\overline{PS} = 4$; adjacent sides $\perp$ and not $\not\cong \left(\sqrt{17} \neq \sqrt{68}\right)$

18. trapezoid; slope of $\overline{PQ}$ = slope of $\overline{RS}$ = undefined; slope of $\overline{SP} \neq$ slope of $\overline{RQ}$ $\left(\frac{2}{3} \neq -2\right)$

19. $(-5, 5.5)$, $(-8, 2)$, $(-7, -2.5)$, $(-4, 1)$; parallelogram

Reteaching with Practice

1. $AB = BC = CD = DA = 3$

2. Slope of $\overline{EF}$ = slope of $\overline{GH} = \frac{1}{4}$; slope of $\overline{FG}$ = slope of $\overline{HE} = -\frac{1}{4}$; so $EFGH$ is a parallelogram; slope of $\overline{EG} = 0$ and slope of $\overline{FH}$ is undefined, so they are perpendicular.

3. kite **4.** isosceles trapezoid **5.** rectangle

Cooperative Learning Activity

1. rhombus, parallelogram **2.** circle

Real-Life Application

1. WY, ME, SD **2.** ME **3.** SD, WY
4. none **5.** WA, ND, TN, IN, VT, CT, KS

6. KY

Challenge: Skills and Applications

1. rhombus; *Sample answer:*

Statements	Reasons
1. Draw $\overline{AC}$.	1. Through any 2 points there is exactly 1 line.
2. $\overline{AB} \cong \overline{BC}$, $\overline{CD} \cong \overline{DA}$	2. Given
3. $\angle ACB \cong \angle CAB$, $\angle ACD \cong \angle CAD$	3. Base Angles Theorem
4. $\overline{AB} \parallel \overline{CD}$	4. Given
5. $\angle CAB \cong \angle ACD$	5. Alternate Interior Angles Thm.
6. $\angle ACB \cong \angle CAD$	6. Transitive Prop. of Congruence
7. $\overline{AC} \cong \overline{AC}$	7. Reflexive Prop. of Congruence
8. $\triangle ACB \cong \triangle CAD$	8. ASA Congruence Postulate
9. $\overline{AB} \cong \overline{CD}$	9. Corresp. parts of $\cong \triangle$s are $\cong$.
10. $\overline{AB} \cong \overline{BC} \cong \overline{CD} \cong \overline{DA}$	10. Transitive Prop. of Congruence
11. $ABCD$ is a rhombus.	11. Rhombus Corollary

2. Square; *Sample answer:* Since $\angle HEF$ and $\angle FGH$ are right angles and are bisected by $\overline{EG}$, $m\angle FEG = m\angle FGE = m\angle HEG = m\angle HGE = 45°$. Since the sum of angles in each of $\triangle FEG$ and $\triangle HEG$ is $180°$, $m\angle F = m\angle H = 90°$. Therefore, $EFGH$ is a quadrilateral with four right angles; by the Rectangle Corollary, $EFGH$ is a rectangle. This implies that $\overline{EF} \cong \overline{HG}$ and $\overline{FG} \cong \overline{HE}$. Furthermore, since $\angle FEG \cong \angle FGE$, we know that $\overline{EF} \cong \overline{FG}$ (Base Angles Theorem). Therefore, $EFGH$ is a quadrilateral with four congruent sides; by the Rhombus Corollary, $EFGH$ is a rhombus. But $EFGH$ is also a rectangle; by the Square Corollary, $EFGH$ is a square.

Lesson 6.6 *continued*

3. *Sample answer:* Draw $\overline{JL}$ and $\overline{KI}$, and let M be the point where they intersect. Since $\overline{IJ} \cong \overline{KL}$ (given), $\angle IJK \cong \angle JKL$ (given), and $\overline{JK} \cong \overline{JK}$ (Reflexive Property of Congruence), $\triangle KJI \cong \triangle JKL$ (SAS Congruence Postulate) Since corresponding parts of congruent triangles are congruent, $\overline{KI} \cong \overline{JL}$ and $\angle 1 \cong \angle 2$, which implies $JM = KM$ (Base Angles Converse). $KM + MI = KI$ and $JM + ML = JL$ (Segment Addition Postulate). Using these addition statements and various properties of equality, we get $MI = KI - KM = JL - KM$ and also $ML = JL - JM = JL - KM$, so $MI = ML$. By the Base Angles Theorem, $\angle 5 \cong \angle 6$. Now vertical angles are congruent, so $\angle 3 \cong \angle 4$. Since $\angle 1$, $\angle 2$, and $\angle 3$ form a triangle and $\angle 4$, $\angle 5$, and $\angle 6$ form a triangle, $m\angle 1 + m\angle 2 + m\angle 3 = m\angle 4 + m\angle 5 + m\angle 6 = 180°$. By the substitution property of equality, $m\angle 2 + m\angle 2 + m\angle 3 = m\angle 3 + m\angle 5 + m\angle 5$ so, using the subtraction and division properties of equality, $m\angle 2 = m\angle 5$. By the Alternate Interior Angles Converse, $\overline{JK} \parallel \overline{IL}$. **4.** *Sample answer:* Using the result from the previous exercise, $\overline{OP} \parallel \overline{RQ}$ and $\overline{OR} \parallel \overline{PQ}$. Therefore, $OPQR$ is a parallelogram, which implies that $\overline{OP} \cong \overline{RQ}$ and $\overline{OR} \cong \overline{PQ}$. But, since $MNPQR$ is a regular pentagon, we also know that $PQ \cong RQ$, so $OPQR$ has four congruent sides. Therefore, $OPQR$ is a rhombus.

5. false; *Sample answer:*

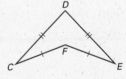

6. true; *Sample answer:* Suppose $GHIJ$ is both a kite and a trapezoid. Then, like the quadrilateral in Exercise 1, $GHIJ$ has two pairs of consecutive congruent sides and a pair of opposite parallel sides, so by Exercise 1, $GHIJ$ is a rhombus. But this contradicts the fact that $GHIJ$ is a kite. So, $GHIJ$ cannot be both a kite and a trapezoid.

7. true; *Sample answer:* A trapezoid contains two pairs of supplementary angles. Each pair contains an acute angle and an obtuse angle, or two right angles. Since at most one of these pairs can contain right angles, the number of acute angles is 1 or 2.

8. false; *Sample answer:*

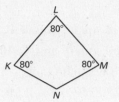

9. false; *Sample answer:*

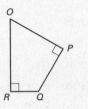

10. false; *Sample answer:*

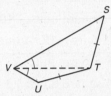

Lesson 6.7

Warm-Up Exercises

1. 16 **2.** 28.26 **3.** 7 **4.** 40 **5.** 9

Daily Homework Quiz

1. parallelogram, rhombus, trapezoid, kite

2. rhombus, square, kite **3.** Kite; $\overline{EF} \cong \overline{FG}$ and $\overline{GH} \cong \overline{EH}$, but opp. sides are not $\cong$.

4. Isosceles trapezoid; $\overline{EH} \parallel \overline{FG}$, and $\overline{EF}$ and $\overline{HG}$ are $\cong$ but not $\parallel$.

Lesson Opener

Allow 10 minutes.

1. 15 square units **2.** 16 square units

3. 15 square units **4.** 8 square units

5. 12 square units **6.** 12 square units

7. 16 square units **8.** 14 square units

Technology Activity

1. *Sample answer:* One diagonal is the perpendicular bisector of the other diagonal.

2. $\frac{1}{2}AC \cdot BE$ **3.** $\frac{1}{2}AC \cdot ED$ **4.** $\frac{1}{2}AC \cdot BD$

Practice A

1. 27 square units **2.** 21 square units

3. 72 square units **4.** 35 square units

5. 99 square units **6.** 110 square units

Answers

Lesson 6.7 *continued*

7. 48 square units   **8.** 64 square units

9. 60 square units   **10.** 24 square units

11. 108 square units   **12.** 60 square units

13. 1800 ft^2 **14.** 1140 ft^2

Practice B

1. 169 square units   **2.** 96 square units

3. 60 square units   **4.** 165 square units

5. 36 square units   **6.** $18\sqrt{10}$ square units

7. 192 square units   **8.** 182 square units

9. 21 square units   **10.** $4\sqrt{5}$ units

11. 12 units   **12.** 4 units

13. 48 square units; 96 square units

14. 24 square units; 72 square units

15. 40 square units; 120 square units

Practice C

1. 98 square units   **2.** $18\sqrt{3}$ square units

3. 135 square units   **4.** 368 square units

5. 72 square units   **6.** 108 square units

7. 85 square units   **8.** 28 square units

9. 312 square units   **10.** 48 square units

11. 40 square units   **12.** 20 square units

13. 14 ft^2 **14.** The area of the kite is 1216 square in. and one square yard is 1296 square inches. So, yes, you have enough material.

Reteaching with Practice

1. 61 sq units   **2.** 81 sq units   **3.** 864 sq units

4. 40 sq units   **5.** 25 sq units   **6.** 98 sq units

Interdisciplinary Application

1. 140 square inches   **2.** 2.25 square inches

3. 179.75 square inches   **4.** 55 square inches

6. 374.75 square inches

Challenge: Skills and Applications

1. 192   **2.** *Sample answer:* (area of $\triangle JKL$) = $\frac{1}{2}b_1h$; (area of $\triangle JLM$) = $\frac{1}{2}b_2h$; (area of $JKLM$) = (area of $\triangle JKL$) + (area of $\triangle JLM$) = $\frac{1}{2}b_1h + \frac{1}{2}b_2h = \frac{1}{2}h(b_1 + b_2)$

3. a. *Sample answer:* Since the diagonals of a parallelogram bisect each other, $OP = OR$ and $OQ = OS$. Since the opposite sides of a parallelogram are congruent, $PQ = RS$ and $QR = PS$. Therefore, by SSS Congruence Postulate, $\triangle OPQ \cong \triangle ORS$ and $\triangle OQR \cong \triangle OSP$, which implies area of $\triangle OPQ$ = area of $\triangle ORS$ and area of $\triangle OQR$ = area of $\triangle OSP$. Furthermore, if h is the distance from Q to $\overleftrightarrow{PR}$, then area of $\triangle OPQ$ = $\frac{1}{2}(OP)h = \frac{1}{2}(OR)h$ = area of $\triangle OQR$. So, all four triangles have the same area.

b. *Sample answer:* Let h be the distance from Q to $\overleftrightarrow{PR}$. Then area of $\triangle OPQ = \frac{1}{2}(OP)h$ and area of $\triangle OQR = \frac{1}{2}(OR)h$. But area of $\triangle OPQ$ = area of $\triangle OQR$, so $OP = OR$. Therefore, $\overline{QS}$ bisects $\overline{PR}$. By a similar argument, $\overline{PR}$ bisects $\overline{QS}$. Since the diagonals bisect each other, $PQRS$ is a parallelogram.

4. 20 square units   **5.** 84 square units

6. $6\sqrt{35}$ square units   **7.** 204 square units

Review and Assessment

Chapter Review Games and Activities

Down

1. 174°   **3.** 3082

5. 360°   **7.** 5264

9. 588   **10.** 240

Across

2. 736   **4.** 400

6. 1052   **8.** 4512

11. 810

Test A

1. polygon   **2.** not a polygon   **3.** polygon

4. not a polygon   **5.** concave   **6.** convex

7. convex   **8.** concave   **9.** sometimes

10. always   **11.** sometimes   **12.** sometimes

13. 9   **14.** 46   **15.** 115   **16.** 2   **17.** no

18. yes   **19.** no   **20.** no   **21.** rhombus

22. trapezoid   **23.** rectangle   **24.** equiangular

25. regular   **26.** none of these   **27.** equilateral

28. 15 m^2 **29.** 35 ft^2 **30.** 12 in.^{2} **31.** 20 cm^2

Review and Assessment *continued*

32. Answers may vary.　**33.** Answers may vary.
34. Answers may vary.　**35.** Answers may vary.

Test B

1. polygon　**2.** not a polygon　**3.** polygon
4. not a polygon　**5.** concave　**6.** convex
7. convex　**8.** concave　**9.** sometimes
10. always　**11.** never　**12.** sometimes
13. 30　**14.** 31　**15.** 13　**16.** 39　**17.** no
18. yes　**19.** yes　**20.** no　**21.** trapezoid
22. parallelogram　**23.** rhombus　**24.** regular
25. equiangular　**26.** none of these
27. equilateral　**28.** 24.5 cm²　**29.** 64 in.²
30. 84 m²　**31.** 25 ft²　**32.** Answers may vary.
33. Answers may vary.　**34.** Answers may vary.
35. Answers may vary.

Test C

1. polygon　**2.** not a polygon
3. not a polygon　**4.** polygon　**5.** concave
6. convex　**7.** convex　**8.** concave　**9.** always
10. never　**11.** sometimes　**12.** sometimes
13. 20　**14.** 39　**15.** 114.5°　**16.** 45　**17.** no
18. yes　**19.** no　**20.** yes　**21.** yes　**22.** yes
23. rectangle　**24.** isosceles trapezoid　**25.** kite
26. regular　**27.** equiangular　**28.** none of these
29. equilateral　**30.** $79\frac{3}{4}$ cm²　**31.** 36 m²
32. 38 in.²　**33.** 25 ft²　**34.** Answers may vary.
35. Answers may vary.　**36.** Answers may vary.
37. Answers may vary.

SAT/ACT Chapter Test

1. E　**2.** B　**3.** C　**4.** E　**5.** C　**6.** B　**7.** A
8. A　**9.** E

Alternative Assessment

1. Complete answers should include:

Property		Rectangle	Rhombus
Diagonals are ≅		yes	
All sides are ≅			yes
Diagonals are ⊥			yes
All ⩭ are ≅		yes	
Exactly 1 pair of opp. sides are ‖			
Exactly 1 pair of opp. ⩭ are ≅			

Property	Square	Kite	Trapezoid
Diagonals are ≅	yes		
All sides are ≅	yes		
Diagonals are ⊥	yes	yes	
All ⩭ are ≅	yes		
Exactly 1 pair of opp. sides are ‖			yes
Exactly 1 pair of opp. ⩭ are ≅		yes	

2. a. 90°　**b.** 43°　**c.** 43°　**d.** 137°
e. 24 units　**f.** 48 units　**g.** 16 units
h. 24 units　**i.** 16 units　**3. a.** 36 units

b. 594 square units　**4.** Answers may vary.
Answers should include five of the six ways to
prove a shape is a parallelogram.

Project: "Puzzling" Shapes

1. a. They are congruent.　**b.** They are congruent, with length equal to one-half the length of the
side of the entire square.　**c.** Its length is one-half
the length of the side of the entire square.
d. They are congruent, with length equal to one-
fourth the length of the square's diagonal.

3. Triangle; *Sample drawing:*

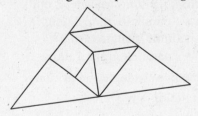

Review and Assessment *continued*

The polygon with the largest number of sides depends on student's creativity. Students should be able to find something with more than ten sides if concave polygons are considered. Check that students have named their polygons correctly.

4. Answers may vary. Sample drawings are given for parts (a)–(d). Note that in the sample drawings triangles have been identified as small (S), medium (M), or large (L).

a. A square cannot be formed with 6 pieces;

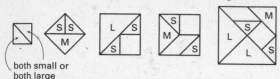

b. A rectangle that is not a square cannot be formed with 2 pieces;

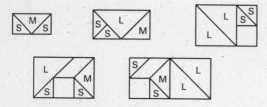

c. A parallelogram that is not a rectangle cannot be formed with 6 pieces;

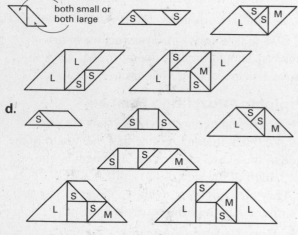

d.

5. a. square, 2 units²; medium triangle, 2 units²; large triangle, 4 units²; parallelogram, 2 units²
b. square, 1 unit²; medium triangle, 1 unit²; large triangle, 2 units², parallelogram, 1 unit²
c. square, $\frac{1}{2}$ unit²; medium triangle, $\frac{1}{2}$ unit²; large triangle, 1 unit²; parallelogram, $\frac{1}{2}$ unit²

d. small triangle, $\frac{1}{2}$ unit²; medium triangle, 1 unit²; large triangle, 2 units²; parallelogram, 1 unit²   **e.** small triangle, $\frac{1}{4}$ unit²; medium triangle, $\frac{1}{2}$ unit²; large triangle, 1 unit², parallelogram, $\frac{1}{2}$ unit².   **f.** small triangle, $\frac{1}{8}$ unit²; medium triangle, $\frac{1}{4}$ unit²; large triangle, $\frac{1}{2}$ unit²; parallelogram, $\frac{1}{4}$ unit²   **g.** small square, $\frac{1}{8}$ unit²; small triangle, $\frac{1}{16}$ unit²; medium triangle $\frac{1}{8}$ unit²; large triangle, $\frac{1}{4}$ unit²; parallelogram, $\frac{1}{8}$ unit²   **h.** This is the same as Ex. 5a where the small triangles have area of 1 unit².

Cumulative Review

1. 46 in., 120 in.²   **2.** 30 ft, 30 ft²

3. If I live in Texas, then I live in Dallas; If I do not live in Dallas, then I do not live in Texas.

4. If $\angle C$ is an obtuse angle, then $m\angle C = 140°$; If $m\angle C \neq 140°$, then $\angle C$ is not an obtuse angle.

5. 27   **6.** 7

7.

Statements	Reasons
1. $\overline{IE} \perp \overline{NC}$	**1.** Given
2. $\angle IEN$ is a right $\angle$	**2.** Definition of $\perp$ lines
3. $\angle IEC$ is a right $\angle$	**3.** Definition of $\perp$ lines
4. $\triangle NEI$ and $\triangle CEI$ are right $\triangle$'s.	**4.** Definition of right $\triangle$
5. $\overline{IE} \cong \overline{IE}$	**5.** Reflexive Equality Theorem
6. $\angle N \cong \angle C$	**6.** Given
7. $\overline{NI} \cong \overline{CI}$	**7.** Converse of Base $\angle$ Theorem
8. $\triangle NEI \cong \triangle CEI$	**8.** HL Congruence Theorem

8.

Statements	Reasons
1. $\angle TIG \cong \angle EGI$	**1.** Given
2. $\angle EIG \cong \angle TGI$	**2.** Given
3. $\overline{IG} \cong \overline{IG}$	**3.** Reflexive Equality Theorem
4. $\triangle TIG \cong \triangle EGI$	**4.** ASA Congruence Postulate
5. $\overline{TI} = \overline{EG}$	**5.** Corresp. parts of $\cong \triangle$'s $\cong$.

Review and Assessment *continued*

9. $\overline{AB}, \overline{AC}$ **10.** $\overline{AC}, \overline{AB}$ **11.** 10 **12.** 15

13. 12 **14.** 1 **15.** $x = 1, y = 2$

16. Since $\angle ADB \cong \angle CBD$, $\overline{AD} \parallel \overline{CD}$ by the Converse of Alt. Int. $\angle$ Theorem. *ABCD* is a parallelogram because both pairs of opposite sides are parallel. **17.** Since $\angle BDC \cong \angle DBA$, $\overline{AB} \parallel \overline{CD}$ by the Converse of Alt. Int. $\angle$ Theorem. *ABCD* is a parallelogram because one pair of opposite sides are both $\parallel$ and $\cong$.

18. $x = 5, y = 4$ **19.** $x = 2, y = 3$ **20.** 13

21. 19 **22.** 156 cm^2 **23.** 120 in.^{2}